Climate Events and Disaster Mitigation
from Policy to Practice

About the Centre

The Centre for Science and Technology of the Non-Aligned and Other Developing Countries (NAM S&T Centre) is an inter-governmental organisation with a membership of 48 countries spread over Asia, Africa, Middle East and Latin America. Besides this, 12 S&T agencies and academic/research institutions of Bolivia, Botswana, Brazil, India, Nigeria and Turkey are the members of the S&T-Industry Network of the Centre. The Centre was set up in 1989 to promote South-South cooperation through mutually beneficial partnerships among scientists and technologists and scientific organisations in developing countries. It implements a variety of programmes including international workshops, meetings, roundtables, training courses and collaborative projects and brings out scientific publications, including a quarterly Newsletter. It is also implementing 7 Fellowship schemes, namely, NAM S&T Centre Research Fellowship, Joint NAM S&T Centre – ICCBS Karachi Fellowship, Joint CSIR/CFTRI (Diamond Jubilee) - NAM S&T Centre Fellowship, Joint NAM S&T Centre – ZMT Bremen Fellowship, Research Training Fellowship for Developing Country Scientists (RTF-DCS), NAM S&T Centre – U2ACN2 Research Associateship in Nanosciences and Nanotechnology and Joint NAM S&T Centre – DST (South Africa) Training Fellowship on Minerals Processing and Beneficiation in Indian institutions. These activities provide, among others, the opportunity for scientist-to-scientist contact and interaction, training and expert assistance, familiarising the scientific community on the latest developments and techniques in the subject areas, and identification of technologies for transfer between member countries. The Centre has so far brought out 73 publications and has organised 101 international workshops and training programmes.

For further details, please visit www.namstct.org or write to the Director General, NAM S&T Centre, Core 6A, 2nd Floor, India Habitat Centre, Lodhi Road, New Delhi-110003, India (Phone: +91-11-24645134/24644974; Fax: +91-11-24644973; E-mail: namstcentre@gmail.com; namstct@bol.net.in).

Climate Events and Disaster Mitigation
from Policy to Practice

— Editors —

Alejandro Linayo Rivero

Jayant K. Routray

Biswajeet Pradhan

CENTRE FOR SCIENCE & TECHNOLOGY OF THE
NON-ALIGNED AND OTHER DEVELOPING COUNTRIES
(NAM S&T CENTRE)

2018
DAYA PUBLISHING HOUSE®
A Division of
ASTRAL INTERNATIONAL PVT. LTD.
New Delhi – 110 002

Publisher's Note:

Cataloging in Publication Data--DK
Courtesy: D.K. Agencies (P) Ltd. <docinfo@dkagencies.com>

International Workshop on 'Mitigation of Disasters Due to Severe Climate Events: From Policy to Practice' (2016 : Colombo, Sri Lanka)
Climate events and disaster mitigation : from policy to practice / editors, Alejandro Linayo Rivero, Jayant K. Routray, Biswajeet Pradhan.
     pages cm
Includes index.
   ISBN: 9789387057272 (International Edition)
     1. Climatic changes--Developing countries--Congresses. 2. Natural disasters--Developing countries--Congresses. 3. Emergency management--Government policy--Developing countries--Congresses. I. Rivero, Alejandro Linayo, editor. II. Routray, Jayant K., editor. III. Pradhan, Biswajeet, editor. IV. Centre for Science and Technology of the Non-Aligned and Other Developing Countries, organizer. V. National Science and Technology Commission (Sri Lanka), organizer. VI. Title.
    LCC QC903.2.D44I58 2016 | DDC 577.22   23

Centre for Science and Technology of the Non-Aligned and Other Developing Countries (NAM S&T Centre)
Core-6A, 2nd Floor, India Habitat Centre, Lodhi Road,
New Delhi-110 003 (India)
Phone: +91-11-24644974, 24645134, Fax: +91-11-24644973
E-mail: namstct@gmail.com
Website: www.namstct.org

Published by   :   **Daya Publishing House®**
                     A Division of
                     Astral International Pvt. Ltd.
                     – ISO 9001:2008 Certified Company –
                     4736/23, Ansari Road, Darya Ganj
                     New Delhi-110 002
                     Ph. 011-43549197, 23278134
                     E-mail: info@astralint.com
                     Website: www.astralint.com

Foreword

Most countries, but especially those located on tropical islands, are highly susceptible to the impacts of severe weather events. Their vulnerability is complex, arising from their specific combination of geographical, geological and socio-economic characteristics. If we are to minimize the risk of disaster due to severe weather events in a given region, there is a need to study and understand the complex contributing factors, many of which are dynamic in time and space, to allow a suitable risk mitigation plan to be developed. It is only when an understanding of the geological, meteorological, sociological, technical and economic aspects of the potential disaster has been gained that the disaster prevention and risk reduction processes can be undertaken effectively, noting of course that this is in itself a challenging set of activities.

Over the past decade, many countries have developed national-level natural disaster prevention policies, in which severe weather hazards feature prominently. However, notwithstanding the value of these policies, the realisation of actual climatic and geophysical events have seen outcomes that have deviated significantly from those anticipated. The reasons are complex – in some cases the hazard event itself has been under-estimated or misinterpreted (for example, failing to recognise the effects of heavy rainfall alongside strong winds during tropical cyclones); in others the capacity of communities to respond, for example through self-evacuation, has been over-estimated. In almost all cases, the response of the authorities has been less effective than had been hoped, usually at no fault of the individuals concerned, for example because communications broke-down early in the event. Whilst various factors are cited as reasons for these failures, work to address these issues is yet to be started in many countries.

The ultimate objective of disaster prevention policies is to minimize human injuries and deaths, and socio-economic losses. A national disaster prevention policy is the vision of a country to prevent, or at least reduce, disaster. However, such policies can only be achieved through multiple parallel approaches that deploy guidelines, action plans and implementation strategies at national, regional and local levels, augmented by a proper hierarchy of risk prevention and management mechanisms. Sadly, many countries lack the capacityto enact and coordinate these essential elements of a disaster prevention policy.

Thus, there has been a growing need for a solid platform to educate, train and network individuals and organisations in developing countries to work beyond existing disaster management policies. I am delighted that the NAM S&T Centre endeavoured to address these issues by organising an international workshop on 'Mitigation of Disasters due to Severe Climate Events: From Policy to Practice' in Colombo, Sri Lanka from 10th to 13thMarch 2016, which was attended by a large number of researchers, experts and professionals engaged in R and D, policy-making and implementation, as well as social activists and other stakeholders from 19 developing countries. This book, which is a follow-up to that workshop and an outcome of the brainstorming that took place during the event, comprises 17 scientific and technical papers by experts from 15 countries.

I applaud the NAM S&T Centre for producing this highly valuable publication, which will be an asset to member states of the NAM S&T Centre, as well as other developing countries engaged in the welfare of the society.

Prof David Petley
Pro-Vice-Chancellor for Research and Innovation
University of Sheffield, U.K.

Preface

Climate change is an accepted and continuing phenomenon felt globally, regionally and also nationally. The impacts are well observed and being measured in different parts of the globe with wide variation in local situations. Climate induced disasters such as temperature rise and its severity at places, flood, drought, cyclone, sea surge and allied activities are growing in numbers and intensities year by year. It is a growing challenge for the scientific communities and practicing professionals working in the forefront to address and provide mitigation as well as adaptive solutions in this regard in order to reduce climate-induced risks. Asian and African countries are not free from the climate change phenomena and induced impacts. The developing Afro-Asian nations have more challenges and constraints to address the impacts effectively. With this back drop, NAM S&T Centre provided a platform in Colombo from March 10-13, 2016 to address issues under a theme "Mitigation of Disasters due to Severe Climate Events: From Policy to Practice" and facilitated the Afro-Asian, European and Latin American professionals to share their works, experience and knowledge in the context of mitigating climate induced disaster risks. The current publication is the product of some selected presentations after careful peer review process.

This edited volume contains 17 articles, which are categorized into 5 sections namely Climate Change, Natural Disasters, Disaster Management Practices and Policies, Technology for Disaster Risk Reduction, and Education and Research.

Climate Change

This section has 5 papers pertaining factors contributing to climate change, and mitigation measures adopted by various countries. The firstpaper "**Changing Weather and Air Quality Patterns in Central European and Mediterranean Region**", by László Bozó presents the effects of global climate change at the regional level and the variability of the local climate using meteorological datasets. The author has discussed the most important elements of mitigation and adaptation-to-climate-change strategy in connection with severe climate events for Central and Mediterranean region. This region is recognized as the most vulnerable zone in Europe because the weather has been largely influenced by long-range atmospheric patterns, which have more frequently shifted the paths of cyclones towards northern direction. The frontal pattern and waving fronts initiate heavy amount of precipitation in Central Europe. However, in recent years, the region is gradually experiencing a decrease in precipitation due to intensive dry downward circulation of Saharan air masses. Which then is decreasing the frequency of Mediterranean cyclones causing long-term drought period. The chapter further deliberates the resultant effect of this phenomenon, which is beginning to be evident in risking human health, which calls for effective adaptation.

The paper on **"Global Warming and Climate Change due to Poor Solid-Waste Management: The Growing Concern in Indian Metropolitans"**, by B.C. Prabhakar and K.N. Radhika focuses on massive greenhouse gas emissions from the municipal solid waste landfills in Indian metropolitans, which is ultimately contributing to climate change. Research uses descriptive approach to discuss policies on municipal solid waste management and climate change in India. The paper has identified unplanned urbanization and poor solid waste management as probable causes. It recommends that municipal solid waste should be regarded as a national issue for ensuring clean environment, public health, and above all preventing global warming.

The paper, **"The Impact of Climate Change on Natural Disasters in Iran"** by Madhi Faravani, Reza Aghnoum, Ali Akbar Moayedi, and Muhammad Ehsan Elahi discusses the overall impact of the vicious effects of human interference with natural climatic setting due to the socio-economic activities in Iran. Human beings in the quest for advancement and better living standard have excessively interfered with the nature. This interference has and will continue to have negative impacts on the environment and social wellbeing of the citizen. In the last decade, Iran has experienced severe drought at a frightening level that is generating a lot of concerns about the country's capacity to increase food production, particularly the cereals. And this is expected to worsen with increasing number of disasters and implications forcrisis management.

The paper on **"Challenges in the Implementation of Climate Change Policy of Nepal"** by Jiba Raj Pokharel tries to highlight the challenges that are likely to be faced in the execution of climate change policies in Nepal. The paper reviews various efforts in formulating climate change policy. Climate change adaptation programs have been reviewed,which is centered on clean energy use, and increasing climate change resilience. It has been found that the political regime is not determined for achieving successful climate change policy implementation.

The paper on **"Mitigating the Effects of Adverse Climatic Conditions in Zambia"** by Shadreck Mpanga, has discussed extensively the approaches, policies and practices by the Zambian authorities to mitigate the effects of the adverse climatic conditions with a focus on the energy and agricultural sectors. The country has witnessed one of its worst summer seasons in 2015. It has impacted considerably, like the loss of its surface water resources due to evaporation and consequently leading to droughts. This phenomenon has forced to devisealternative approaches to sustain those key sectors of the economy.

Natural Disasters

This section discusses papers on natural hazards specially on flooding, and initiatives taken by disaster management institutions to cope against disasters. The first paper in this section on **"Natural Disasters and Climate Change in Nigeria: Issues and Mitigation Strategies"** by Agoro Playiwola discusses the myriad of direct impacts of climate change and related disasters in Nigeria and the various intervention policies and strategies to mitigate them. Like other parts of the world, the Sub-Sahara Africa, particularly Nigeria is also contending with the ideatominimize the effect of climate change such as drought, extreme temperature, floods, landslides, and so on. A particularly disturbing disaster in focus is the debilitating erosion due to floods of Nanka community in Anambra State that has resulted in a damageworth of millions of dollar with no proper solution on sight.

This section has five papers, which discuss the policies at the national or at the regional level, and subsequently the disaster management practices attuned to the policies. The paper on **"Flash Flood Assessment of 14ᵗʰ November 2014 in Colombo due to Short Period High Intense Rainfall"** by A.R. Warnasooriya, A.C.M. Rodrigo, and K.H.M.S. Premalal focuses on the assessment of flash flood caused by intense rainfall of 56mm over the course of a short time period of 20 minutes in Colombo from November 2014. For this investigation, the authors used the data from meteorological department and NOAA NCEP. The data was put into the Weather Forecast Model version 3.7.1 with 3DVAR data assimilation. In addition to this satellite images and atmospheric water content was also analyzed as a part of this research. The results of the analysis in this papers throw light on the reason of this flash flood being associated with dry winds meeting with the moist winds. It also attributes the flash floods to the uncontrolled growth of the city complimented by illegal construction and the collapse of the natural drainage system. This study can be useful to alert people in the danger areas by monitoring the dry winds and the moist winds meeting each other (Dry Line) in case of future possibilities where this kind of event can take place.

The paper on **"Children's Experiences, Participation and Resillience to Flooding: Insights from Muzarabani, Zimbabwe"** by Chipo Mudavanhu, explores parents and stakeholders' views on children's participation in flood risk reduction in Muzarabani, Zimbabwe. The study used the qualitative approach to document parents' and stakeholders' views about children's participation in disaster risk reduction (DRR). Results indicated that most parents do not consult children during emergencies. The study recommends for the creation of school-community linkages, developing institutional capacity to deal with disasters and strengthening school physical conditions.

Disaster Management Practices and Policies

The paper on, **"Disaster Management Practices in Cambodia"** by Viseth Ung and Leng Heng An portrays the practices involved for disaster management in Cambodia especially to the policymakers, academia, researchers, students, and other interested stakeholders for their own use further. Primary and secondary data seem to be used for this study with accounts of the total losses from recent disasters in Cambodia. This paper deals with the evolution to disaster management in Cambodia, the plans, policies, strategy and framework for Disaster Management, the amount of losses from the flood disaster of 2011, the efforts taken by the government and lastly the recent major projects related to disaster management in the country. Although, the plans, policies and the frameworkare in place but it was observed that there is a general lack of appreciation regarding the efforts taken by the disaster risk management institutions, along with lack of understanding of maintaining a good database for decision making, inadequate resources (financial and human) for disaster risk management activities and understanding among different stakeholders.

The paper on **"A Synopsis of Disaster Management in South Africa - An Intergovernmental Approach to Disaster Prevention, Mitigation and Response"**, by Dumisani Emmanuel Mthembu analyses the policy and regulatory framework for disaster management in South Africa. It discusses on the evolution of disaster management approach in South Africa, where there has been decentralizing and institutionalizing disaster management in all spheres of government. This paper is based on secondary data and follows document analysis method. The author has examined the data from the various documents reviewed. The documents evaluated are specific to disaster and disaster management in South Africa.This paperrecognizes that the policies are in place, but the implementation of policies and capacities of institutions at the local level has not been realized. This is an issue faced by many countries in the developing world and their needs to be a consorted effort towards understanding that needs to be done to build capacities at the local level.

The paper on **Possible early warning for landslide in Sri Lanka using "Antecedent Daily Rainfall Index": A case study of Meeriyabedda Landslide on 29th October 2014** by W.N.S. Rupasinghe and K.H.M.S. Premalal is a study carried out to find a suitable method for identify possibility of early warning using antecedent daily rainfall index. Antecedent daily rainfall indexwas applied for 5 landslides which occurred in Badulla district and the constant in the equation for antecedent daily rainfall index was identified as 0.9. The results gave good indication for landslide with the existing rainy condition even few days early that the event occurred.

The paper on **"Disaster Mitigation Policies and Practices in South Asia with focus on Social Capital"** by Jayant K. Routray and Saswata Sanyal is aimed towards understanding the disaster mitigation policies and practices of the countries in South Asian region and trying to understand if the role of social capital has been appreciated in them. The data was collected from journal papers, articles, websites, and reports using the given keywords: climate change, extreme events, climate-induced disasters, disaster risk reduction, disaster mitigation policies and practices and social capital. It was understood from the disaster mitigation policies from different countries in the South Asian region that Community Based Disaster Risk Management (CBDRM) has a prominent place in the national disaster management frameworks of all the countries. There is hardly any recognition given to the invisible resource: social capital that every community possesses and which can be used for designing and implementing CBDRM programs. Social capital has a vital role to play in disaster mitigation, by helping the community to cope with stress and helps to mitigate adverse effects of hazards, by using the bonding between community members and the connections they have with people outside their community. It also acts as a form of informal insurance for the community till the time the outside help comes after a disaster. Advantages that social capital poses should be taken into consideration by the policy makers while planning for disaster mitigation and preparedness in the future, especially when planning about community-based initiatives where social capital is inherent and can be mobilized with ease.

The paper on "**Existing National Policies on Natural Disaster Management and Implementation: A Case of Landslides and Mudslides in Uganda**", by Wilberforce Kisamba Mugerwa, presents elaborate appraisal of national policies on natural disaster management and implementation in Uganda. The author used landslides and mudslides from the Mt. Elgon areas in eastern Uganda as a case study. He tried to provide a holistic view on the institutional arrangement for disaster mitigation effortslike, the challenges faced during implementation of the policy. Furthermore, the chapter provides suggestion for improving the existing policy ondisaster, which is relevant to landslides and mudslides. This can be done by sensitizing the institutions, policy makers, government and stakeholders on the danger of neglecting disaster preparedness and emergency management and the threat it can pose to the prosperity of the country and the citizens.

Technology for Disaster Risk Reduction

This section deals with technology in disasters. This section consists of two papers. The first paper on "**Identification of Indonesian Technology Readiness in Disaster Risk Reduction**", by Adawiah, Ophirtus Sumule, and Irsan Pawennei tries to identify the disaster-related technologies and provide policy recommendations related to the utilization of innovative products on disaster risk reduction that are used or needed in Indonesia to reduce the wrath of disasters in the country. Analyzing the technology that can lead to innovation with regards to disaster readiness, and at last analyzing the current policy with regards to the technology for disasters is in place in Indonesia. Out of the 50 technology products discussed in the literature review section, as many as 37 technology products are solely used for the prevention and preparedness of disasters. The stakeholders involved in this study select 12 technology products out of 50 that need to be further developed among other technologies mentioned. It is also mentioned that these technologies should accommodate local wisdom. According to the study all the stakeholders need to work together for making these products successful.This kind of approach can be followed collaborating with other technology providers and countries in the region for sharing of information and technology.

The paper on "**Confiability Analysis of the Automatic Weather Stations Network of Sri Lanka**" by Nuwan Kumarasinghe discusses the role of technology and the unexpected environmental challenges encountered in the automatic weather station networks. This study also focuses on the functionality of the sensors in Automatic Weather Stations.This study is mainly concentrated on the performance of AWS installed at 21 meteorological stations. The paper discusses about the different sensors present in the automatic weather station. It tries to understand the behavior and performance of the different sensors from 2009-2015. According to the paper the most critical sensor in the Automatic Weather Stations of Sri Lanka is the ultrasonic wind sensor. The paper also finds out that the VSAT at Galle, Jaffna and Pollonnaruwa performs at less than 50 per cent when compared to other sites. The faults in the Automatic Weather Stations have been categorized, among which are obstruction of the antenna path by trees or other structures, corrosion of the sensors at the coastal weather stations, insects inside the data logger, signal interference, and bird strikes. The author suggests that it is important to understand the causes of the faults in sensors at the Automatic Weather Stations across different places in Sri Lanka, as it can help in the future while designing new systems so that these faults are not repeated.

Education and Research

This is the last section of the volume consisting of two papers. The paper on "Academic Hubs: Using Applied Research and Community Services to Build Resilience of Nations and Communities to Seismic Disasters", by Jalal Al-Dabbeek, Hatim Alwahsh, Sami Sader, Abdel Hakeem Juhari, Barbara Borzi, Fabio Germagnoli, Paola Ceresa, and Ricardo Monteiro illustrates An-Najah National University's Urban Planning and Disaster Risk Reduction Center (UPDRRC) efforts to increase the disaster resilience of the Palestinians. This was done by bringing together different stakeholders and using the scientific knowledge and community services driving the change in policy, preparedness efforts and public awareness. The paper explains the role of different academic hubs using science and technology to build resilience, and support action for strengthening Palestine`s capacity for seismic risk mitigation using different activities like trainings and workshops. These hubs help in increased awareness of seismic risk among different stakeholders, sharing database dealing vulnerability data and making guidelines for risk management policies and reducing vulnerability. According to the paper these kinds of hubs can not only be replicated in Palestine, but also in other developing countries for better understanding of risk.

The paper on, "A Designing Experience of a National Research Agenda on Disaster Risk Reduction", by Alejandro Linayo focuses on the experience of developing and implementation of the National Research Agenda for Disaster Risk Reduction in Venezuela between 2000-07 by Ministry of Science and Technology. Other than this the goal of this research was also to promote a deeper commitment of the scientific and academic community towards improving the disaster risk reduction capacities. The paper discusses the evolution of the National Research Agenda from the scratch to its current form now. In the process it also illustrates the methodological framework adopted by the Ministry of Science and Technology for developing the agenda and subsequently the implementation and results of the program. This program supported about 55 projects worth about $ 4.5 million conducted by scientists and academicians. During the course of implementation of the agenda it was realized that it is vital to include disaster risk reduction in the education system of the countries and universities for better training and education of the people of the country towards reducing risk. The paper recommends to have projects which include aspects such as politics, institutions, social and cultural conditions to further define vulnerability of disasters. This can help country's decision makers to better understand the different nuances of disaster risk reduction.

The objective of this publication is to present a group of selected experiences on applied science and technology efforts to address disaster risk reduction issues. Different chapters of this publication present useful information and valuable lessons to illustrate some important advances to promote the roll of science and technology in disaster risk reduction with some future challenges to be addressed while reinforcing the role of science and technology for a less rhetoric and more practice oriented disaster risk reduction policies.

This publication also is expected to enable readers to understand the wide range of aspects linked with integrated disaster risk reduction. It is topic that is in high demand today across the world to the development actors, and particularly to the academicians. Consequently, it is crucial to understand that effective integrated disaster risk reduction and a safer futurecan only be possible if we clearly comprehend how we must improve our response capacities after a disaster, and also how to we shouldreduce the social vulnerabilities of the people living in the high-risk areas.

The challenges to promote a more prospective and corrective approach of disaster risk reduction motivates us today to think how can we promote more and better sectorial investments and sustainable development models. It is beyond doubt, how important is to continue our efforts in preparedness and response technologies. Consequently, it is also important to understand that these effortsin our development models will never be enough until and unless we can comprehend the challenges that act as barriers to reduce the risk of natural disasters in a community.

From this point of view, this publication is strongly recommended not only for first responders or scientists/academics focused on the study of natural events such as earthquakes, landslides, volcanoes, cyclones, *etc.*, that trigger disaster scenarios, but also for scientists and professionalsengaged in economic, urban and rural planning, education, social sciences, engineering, architecture, public health, and a wide range of additional areas of professional specialization.

Prof. Dr. Alejandro Linayo Rivero

Prof. Dr. Jayant K. Routray

Prof. Dr. Biswajeet Pradhan

Introduction

Climate Change' is a matter of great concern at the global level since it is leading to causing weather-related disasters and in making communities more vulnerable to the effects of disasters. Disaster is a serious disruption of the functioning of a community involving widespread human, material, economic or environmental losses and impacts that exceed its ability to cope using its own resources. Not only social, economic, and geographic factors influence vulnerability and exposure but governance also plays a crucial role.

The susceptibility of a country to climate change is closely linked to its geological, geographical and socio-economic character. Climate change affects all countries, but people in the developing countries, especially LDCs, and poor people in rich countries are more likely to suffer the most owing to intrinsic vulnerabilities to hazards and comparatively low capacities for risk reduction measures. This needs disaster mitigation measures to reduce or if possible, totally eliminate the impacts and risks of hazards through proactive measures taken before an emergency or disaster occurs. An approach to disaster risk management will aim to decrease the vulnerability by adopting prevention and mitigation measures to reduce the physical impact and to increase the coping capacity and preparedness of the community, in addition to providing traditional emergency care (response) once the disaster has occurred.

For alleviating the risk of disaster due to severe climate events, there has been a growing need for a concrete platform to educate and train the individuals and entities in the developing world and network with them to act beyond the disaster management policies. Keeping this in view, the NAM S&T Centre in partnership with the National Science and Technology Commission (NASTEC) of Sri Lanka organised an international workshop on 'Mitigation of Disasters due to Severe Climate Events: From Policy to Practice' in Colombo, Sri Lanka during 10-13 March 2016, which brought the scientists, experts and professionals engaged in R and D, policy making and implementation, social activists and other stake holders to a common forum for sharing views and experiences for the development of a road map for reducing the risks in real situations.

64 senior professionals from 19 countries - Cambodia, Egypt, Hungary, India, Indonesia, Iran, Malaysia, Mauritius, Myanmar, Nepal, Nigeria, Pakistan, Palestine, Thailand, Uganda, Venezuela, Zambia and Zimbabwe, and the host country Sri Lanka – attended the workshop.

As a follow up of the Workshop, the present book - **Climate Events and Disaster Mitigation From Policy to Practice** has been edited by Prof. Alejandro Linayo Rivero, Prof. Jayant K. Routray and Prof. Biswajeet Pradhan. There are 17 scientific and technical papers contributed by the experts from 17 countries. The papers in this book have been categorised in five sections, namely Climate Change, Natural Disasters, Disaster Management Practices and Policies, Technology for Disaster Risk Reduction, and Education and Research.

I have my full appreciation for the efforts put in by the editors of this publication on technical editing of the manuscripts. I also acknowledge the interest and valuable efforts of entire team of the NAM S&T Centre and am especially thankful to Dr. (Mrs.) Kavita Mehra, Mr. M. Bandyopadhyay, Ms. Nidhi and Mr. Pankaj Buttan in compiling and checking the manuscripts, liaising with the authors, cover page designing, proof reading, formatting and taking all the necessary actions in giving a shape to this volume.

I am sure this book will be useful to all those associated with mitigation of disasters due to severe climate events, from researchers to policy makers, non government organisations and government officials in the developing countries.

Prof. Dr. Arun P. Kulshreshtha
Director General,
NAM S&T Centre

Contents

NATURAL DISASTERS

DISASTER MANAGEMENT: PRACTICES AND POLICIES

TECHNOLOGY FOR DISASTER RISK REDUCTION

EDUCATION AND RESEARCH

Climate Change

Chapter 1

Changing Weather and Air Quality Patterns in Central European and Mediterranean Region

László Bozó

Member of the Hungarian Academy of Sciences,
Hungarian Meteorological Service
H-1024 Budapest, Kitaibel Pál utca 1., Hungary
E-mail: bozo.l@met.hu

ABSTRACT

Climate variability and the increasing number and intensity of various extreme hydrometeorological events have their significant effects on socio-economic activities and the natural systems. Such a tendency could be detected especially for droughts, floods, heavy rainfalls, heat waves and air pollution episodes during the past decades. Concerning climate change, the Central European and Mediterranean Region belongs to the most vulnerable areas in Europe. Sustainable development is one of the key issues of the next decades. It includes a broad horizon of human activities like water and drought management, energy production, transport, industrial activity, agriculture, or human health care.

Weather in Central and Mediterranean regions of Europe are mainly determined by long-range atmospheric patterns. As a convincing consequence of climate change, the paths of cyclones are shifted more frequently towards Northern direction, only the Southern sections of meteorological fronts are able to pass through these regions. Most amount of precipitation detected in Central Europe is due to these frontal patterns and waving fronts. It is also broadly documented that the convection-initiated early summer excess precipitation amount in Central Europe has a decreasing tendency during the past decades. Due to the intensive dry downward circulation of Saharan air masses, the frequency of Mediterranean cyclones is also decreasing, causing long-term drought periods in Central Europe. The latest extensive dry period was registered in 2011-2012. On the other hand, in 2010, this region experienced two deep Mediterranean cyclones causing heavy rains and serious floods in many countries of South-Eastern and Central parts of Europe. Meteorological datasets are to be presented to characterize the effects of global climate change on regional atmospheric processes.

Besides the physical properties, the chemical composition of the lower atmosphere is also influenced by changing climate. Longer dry and sunny meteorological conditions in summer are highly favourable for the formation of photochemical smog, where surface ozone concentration may exceed the air quality threshold values under these conditions. In winter, in stable anticyclonic situations the weather is calm, which leads to reach very high particulate concentrations in urban areas and in their regional background, causing increasing risks for human health.

Most important elements of mitigation and adaptation-to-climate-change strategy in connection with severe climate events for Central and Mediterranean region are to be outlined and discussed in the presentation.

Keywords: *Climate change, Extreme meteorological events, Mediterranean cyclone, Air quality, Mitigation and adaptation-to-climate-change.*

Introduction

The global climate change, initiated by human activities, has been in the focus of scientific research for several decades. Long time had to be elapsed until politicians and decision makers also started to deal seriously with this issue. From the early 1970's some highly ranked politicians recognized the significance of this hazard and took more seriously the increasing amount of information from the scientific fora on the changes of quantities of greenhouse gases in the atmosphere, the long-term trends of global climate characteristics, and the scenarios derived from the simple early numerical models. At the UN Conference on Human Environment held in Stockholm in 1972 and in the documents adopted by the participants, the risk of the changing climate as caused at least partially by anthropogenic activities was mentioned already in very general terms. The need to cooperate at international level on emerging large-scale environmental problems – that time primarily focusing on the long-range effects of acidification - has been reinforced in the Final Act of the Helsinki Conference in 1975. Complex field experiences and dynamic climatological modelling studies were carried out in the second half of the 1970's in the framework of the interdisciplinary and intergovernmental Global Atmospheric Research Programme (GARP) that was jointly coordinated by the WMO and the ICSU. This programme lasted 15 years and contributed also to strengthening of the cooperation of those scientific centres, which were dealing with various aspects of the climate system (HAS, 2010).

The World Climate Conference (WCC), held in Geneva in 1979, was the first global gathering of scientists devoted expressively to the state and changes of the global climate. The conference reviewed the results gained that far and identified the areas of future research activities and collaboration. It adopted a declaration with a call "to foresee and to prevent potential man-made changes in climate that might be adverse to the well-being of humanity". Hungarian scientists actively contributed to both the cooperative work of the GARP and to activities of this conference. The real breakthrough in the relation between climate change science and climate change policy, *i.e.* the researchers and the policy-makers happened in the 1980's and this resulted in commencement of international cooperation on elaboration and implementation of the response policies. The World Climate Programme (WCP) was launched and several international conferences were held on various aspects of the changing climate. The UN General Assembly in 1987 adopted a resolution on the report of the Brundtland Commission (World Commission on Environment and Development), which *inter alia* emphasized the danger of the anthropogenic change of climate. Upon the initiation of the WMO and the UNEP, the Intergovernmental Panel on Climate Change (IPCC) was established in 1988.

Hungary contributed to and participated also in these events. A special national report was presented to the Second World Climate Conference, and the President of the Republic of Hungary delivered an important statement regarding the need for international cooperation in combating the climate change. Studies were published in Hungary in relation to carbon dioxide emissions to the atmosphere. A Climate Subcommittee of the Meteorological Scientific Committee of the Hungarian Academy of Sciences was established and prepared in 1991 the statement of the Academy on climate change and the necessary actions. In Hungary, preparatory activities started also under the auspices of the ministries of the environment and foreign affairs on the participation in the UN Conference on Environment and Development. As part of this process, comprehensive analyses were completed and published on the risk, causes and potential impacts of climate change, and on the options of the various response measures (HAS, 2010).

Materials and Methods

Climatological Conditions

Climatological data gained and archived in Hungary have been processed by means of statistical homogenization and interpolation. Hungary is situated in Central Europe within the Carpathian Basin, between the 45°45'N and 48°35'N latitudes, about halfway between the Equator and the North Pole, in the temperate climatic zone according to the solar climatic classification. Its climate is very erratic. One of the main reasons for this is the fact that Hungary is situated in between 3 climatic zones: (i) the oceanic climate with less varying temperature and more evenly dispersed precipitation amount; (ii) the continental climate with more extreme temperature and relatively moderate precipitation; (iii) also, a Mediterranean effect with dry weather in summer, and wet one in winter (HMS, 2016).

For a shorter or longer period of time any of these types can become prevailing. Due to these reasons great differences can occur in the weather of the country, despite of its lower altitudes and relatively small extent.

The other main determinant is orography. As the country is situated in the Carpathian Basin - more than half of its surface are plains below 200 metres, and the area above 400 metres is less than 2 per cent - primarily the effect of the Carpathians should be underlined.

Hungary is about halfway from the ocean to the inner parts of the Eurasian continent. In the summer half-year the dominating air masses are of oceanic origin, in the winter mostly continental ones. The NW-SE distribution of the meteorological variables shows the effect of the Atlantic Ocean, while the SW-NE distribution the effect of the Mediterranean Sea.

Hungary is in the conveyor belt of the Westerlies, due to the location of the country - surrounded by the Alps and the Carpathians - the prevailing wind direction is northwestern, while the southern wind has a secondary maximum.

The climate of Hungary can not be classified by using one of the global climate classifications (e.g. Köppen or Trewartha) to adequately describe the differences within the country. We must find another classification method. This could be done based on the work of a Hungarian climatologist György Péczeli, who - taking into account the aridity index and the growing season length - separated 16 climatic zones, from which 12 can be found in Hungary.

Following this classification, greater part of the country has a moderately warm - dry climate. The area of the rivers Körös and Maros, the lower part of the river Danube is warm - dry. The northeastern region of Hungary (Nyírség) is more likely moderately cool - dry, while the nearby Plain of Szatmár is moderately cool - moderately wet.

In the Southern Transdanubia region the moderately warm - moderately dry, the moderately warm - moderately dry, while in the Western Transdanubia the moderately cool - moderately dry, and the moderately cool - moderately wet climate zones are typical. Higher altitudes of our mountains have a cool - moderately dry, and a cool - moderately wet climate, only in the Mountains of Kőszeg near the western borders with Austria we have a cool - wet climate.

Bias-corrected Regional Circulation Models have been applied in order to estimate the future climate characteristics for the next decades. Climate change is often considered as higher temperature values and more frequent heat waves (*e.g.*, Pongrácz *et al.*, 2013). However, it usually involves more intense and more frequent extreme events related to excess or lack of precipitation (*e.g.*, severe dry spells, heavy precipitation, intense thunderstorms), too. It emphasizes the importance of climate research in quantifying the detected past and the projected future changes from global to local spatial scales. Frequent hot weather situations in summer and overall increasingly warm climatic conditions are quite straightforward and broadly documented consequences of global warming. Global and regional warming induced effects on precipitation are not as clear as on temperature, because the higher spatial and temporal variabilities might hide any robust changing signal. Nevertheless, precipitation is one of the most important meteorological variables, since it considerably affects natural ecosystems and cultivated vegetation as well as most of human activities. Extreme precipitation events – both excessive, intense rainfalls and severe droughts – may result in several environmental, agricultural, economical, and natural disasters. The lack of precipitation for extended period and coincidental intense heat wave often lead severe drought events (Pongrácz *et al.*, 2014).

Climate Change and Water Management

Hungary is geographically located in such an area of the Carpathian Basin, which is prone to high risks of flooding, inland excess water inundation and drought. Following a nearly two decades long dryness and drought period, four highly dangerous floods passed through the Hungarian part of the River Tisza in the period 1998-2001. In 2006 there were two simultaneously occurring extreme floods in the Tisza and in the Danube. Due to the extreme weather conditions of the past years there were serious inland excess water damages and other locally occurring inundations. Excess water, a speciality of the Hungarian plains and flat lands, occurs when precipitation or snowmelt water cannot infiltrate into the soil (mostly due to semi-impermeable upper soil layer) and fills the depressions of the flat land. In Hungary the elimination of water-damages became a task of strategic importance in the past years, due to the increasing frequency of extreme events. One would need global or at least catchment basin based thinking for the design of regional and local strategies. Evidently bi- or multi-later international co-operation is needed for such an approach. Flood and excess water fighting/control have great traditions in Hungary. In the past 150 years main-levees of a length of about 4200 km were built along the rivers. The height of the levees (the design flood level) was calculated as the hundred-year flood level (of 1 per cent probability of occurrence) plus 1.0 metre safety height. The objective was to enhance the rapid passing of flood hydrographs without causing damages. In the case of floods exceeding the design flood level the opening of emergency storage reservoirs facilitated the reducing of flood peaks (HAS, 2010).

Flood levels have been considerably increased in the past decades. This was partly due to the weather's becoming more extreme and to the decreasing of the flow carrying capacity of the flood-channel. Anthropogenic impact on the catchment basin also contributed to increasing and flashier runoff. Further rising of the height of flood levees will not provide an efficient means of flood control. Extreme weather events of the past years and the associated floods and droughts unambiguously indicated that the earlier water management practices couldn't be continued. Instead of the approach of "fighting the floods" the concept of "living with the floods" should be followed. Similar changes in the approach are needed in handling excess water and drought problems, which should be aimed at changing land uses and thus reducing the potential damage. Sufficient spaces should be secured for waters and as much water should be stored as possible, diverting these waters to places in water shortage (HAS, 2010).

Results and Discussions

General characteristics of annual mean temperature and precipitation are shown in Figures 1.1. and 1.2 (HMS, 2016). Temperature has a slight increasing trend with record high values in the period of 2007-2009, while the variation of precipitation amount did not show statistical significance during the period investigated.

Looking at the whole last century, it can be stated that the climate has warmed in Hungary, as well (*Figure 1.3*). Based on studies of homogenized data series we

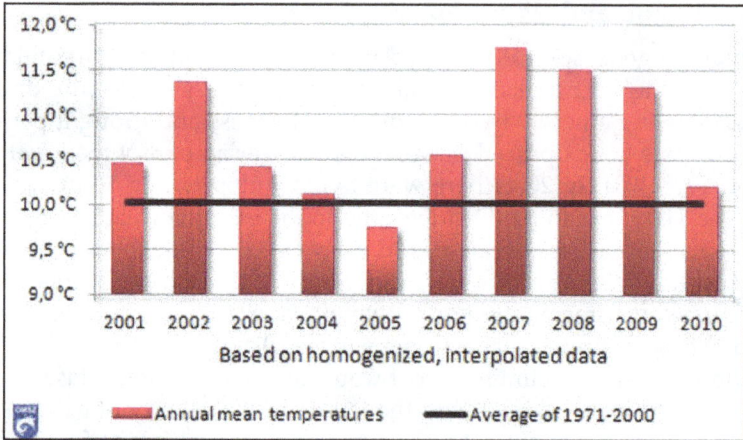

Figure 1.1: Annual Mean Temperatures in Hungary, 2001-2010.

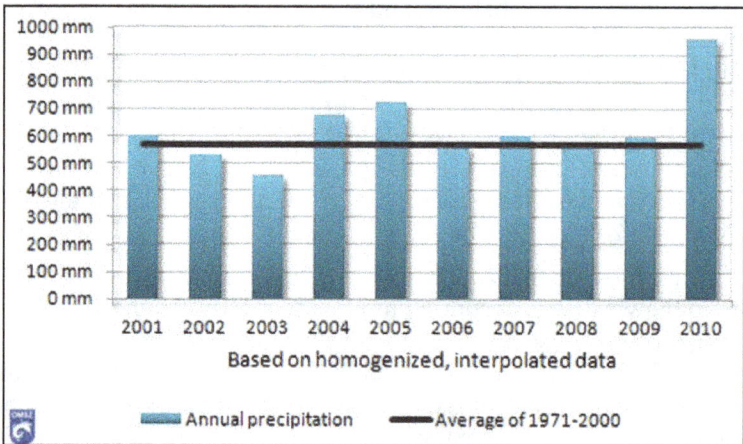

Figure 1.2: Annual Mean Precipitation in Hungary, 2001-2010.

can say that the characteristics of the Hungarian series follow the global tendencies, with a bit greater variability (HMS, 2016).

For instance, in 2003 a long-lasting, devastating heat wave occurred throughout Europe (Stott *et al.*, 2004), causing death of hundreds of people. In Hungary, the year 2003 was generally dry with 17 per cent less annual precipitation than the 1971–2000 average. The Europe-wide heat wave in the summer superposed to these overall dry conditions, resulting in severe drought Pongrácz *et al.*, 2014). The estimated monetary damage in the Hungarian agriculture reached 50–55 billion HUF by the end of the year (HAS, 2010). Another hot and dry summer from the past decades occurred in 2007, this drought resulted in reduced harvest of maize in Hungary and caused at least 80 billion HUF loss (HAS, 2010). On the contrary, in May 2010, the total rainfall in Hungary largely exceeded the average monthly precipitation of the 1971–2000 base period for May, namely, almost three times more precipitation occurred than usual (Móring, 2011). The excessive precipitation led to

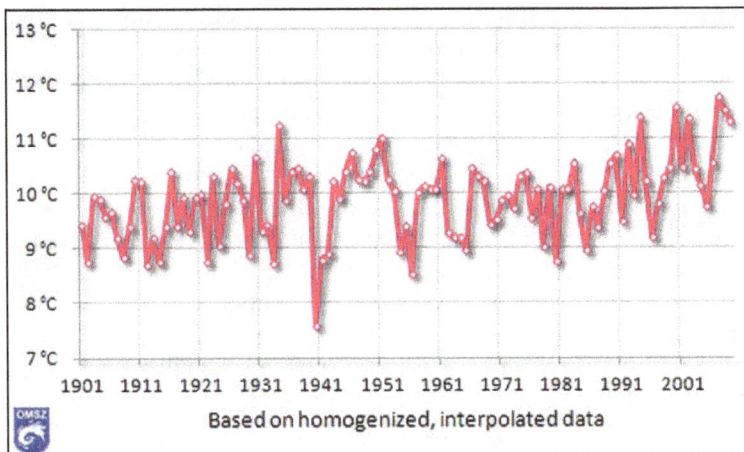

Figure 1.3: Annual Mean Temperature in Hungary, 1901–2009.
(*based on homogenized and interpolated data*).

inland inundation and floods on Sajó, Hernád, Bodrog, and Bódva rivers resulting in more than 10 billion HUF of defence and recovery costs. Overall, the year 2010 became the wettest year in Hungary since 1901 with 959 mm annual precipitation amount exceeding the annual mean of the 1971–2000 period by 65 per cent (Móring, 2011). Besides Hungary, a large majority of the Central/Eastern European region was hit at the same time by severe floods.

According to the 11 bias-corrected Regional Climate Model (RCM) simulations in the 2021–2050 period, smaller changes are projected than in the 2071–2100 period (Pongrácz *et al.*, 2014). By the mid-century, only a few 315 RCM simulations project statistically significant seasonal changes and the average estimated changes do not exceed 11 per cent. In most of the indices, the signs of the projected changes are identical, which implies that the tendencies are likely to continue throughout the 21st century. In general, RR1 and RR5 (precipitation days exceeding 1 mm and 5 mm, respectively) are projected to decrease in summer and increase in winter. However, by the late century, almost all RCM simulations estimate significant decrease in summer (the average projected decrease is 27 per cent relative to the reference period both for RR1 and RR5), and increase in winter for RR5 (the average projected increase is 25 per cent). CDD (*Maximum length of dry spell, i.e., maximum number of consecutive dry days :* $R_{day} < 1\ mm$) and MDS (*Mean length of dry spell :* $R_{day} < 1\ mm$) in summer are projected to increase significantly in Hungary by the end of the 21st century (by 42 per cent and 41 per cent on average, respectively), clearly implying considerably drier future summers (Pongrácz *et al.*, 2014).

On the basis of our present knowledge it is difficult to assess the impacts of global climate changes on the frequency of occurrence of flood events. It is especially difficult to plan the flood control strategies of smaller streams, as their floods are induced mostly by large rainfalls of smaller spatial extent, for which the various climate change models give no forecasts or estimates. In our larger rivers where the high floods occur mostly upon the joint effects of snowmelt and rainfall the peak flood levels may be rising and may occur at earlier point of time than nowadays. The long-term strategies of managing excess waters will not be basically changed by the changing climate. We will have to get prepared for the occurrence of extreme excess water inundations in the end of the winter and early spring also in the future. The control of excess waters will be much more affected by the changes of land use and therefore the planning of excess water control and the development of land uses must be closely harmonised. It is highly probable that climate change will affect the high flows of the rivers to some extent, but it cannot yet be justified that these changes will be dominating. Another two affecting factors are the anthropogenic changes on the catchment and the occurrence of weather situations that have not yet been recorded. Therefore the level of flood protection must be anyway increased. It is highly probable that more and more extreme floods will occur, although their time of occurrence cannot be assessed. Technical and organisational means and their combinations can assure protection against water damages (HAS, 2010).

Increasing number and intensity of flash-floods in Hungary indicates the risks arising from intensive convective meteorological systems which should be taken into account when saving lifes and properties, as well as during the designing of critical infrastructures (Figure 1.4).

Figure 1.4: Flash-Flood in Hungary in June, 2004.

Weather in Central and Mediterranean regions of Europe are mainly determined by long-range atmospheric patterns. As a convincing consequence of climate change, the paths of cyclones are shifted more frequently towards Northern direction, only the Southern sections of meteorological fronts are able to pass through these regions. Most amount of precipitation detected in Central Europe is due to these frontal patterns and waving fronts. It is also broadly documented that the convection-initiated early summer excess precipitation amount in Central Europe has a decreasing tendency during the past decades. Due to the intensive dry downward circulation of Saharan air masses, the frequency of Mediterranean cyclones is also decreasing, causing long-term drought periods in Central Europe. The latest extensive dry period was registered in 2011-2012. On the other hand, in 2010, this region experienced two deep Mediterranean cyclones causing heavy rains and serious floods in many countries of South-Eastern and Central parts of Europe (Figures 1.5 and 1.6).

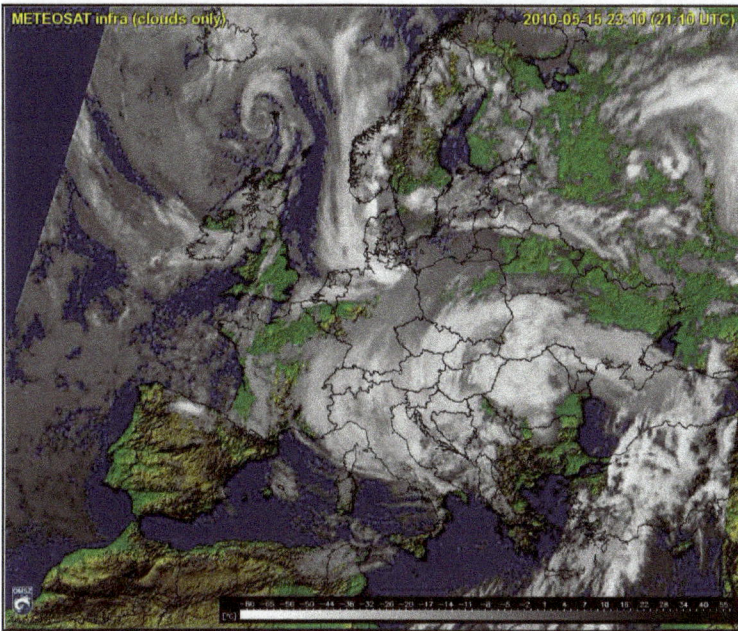

Figure 1.5: Deep Mediterranean Cyclon over Central and South-eastern Europe in May, 2010.

Besides the physical properties, the chemical composition of the lower atmosphere is also influenced by changing climate. Longer dry and sunny meteorological conditions in summer are highly favourable for the formation of photochemical smog, where surface ozone concentration may exceed the air quality threshold values under these conditions. In winter, in stable anticyclonic situations the weather is calm, which leads to reach very high particulate concentrations in urban areas and in their regional background, causing increasing risks for human health and built environment.

Conclusions

Changing climate in Central Europe increases the risk of flooding, drought and inland excess waters. The extent of damages caused is expected to increase and therefore the tasks of the water management in fighting floods, excess waters and droughts must be defined in their interaction and joint causes, putting the main emphasis on supporting multiple effect measures, such as the creation of multipurpose reservoirs or the breaking through the semi-impermeable upper soil layer with various ploughing techniques so as to avoid inland excess water, which is generated mostly due to the lack of appropriate infiltration capacity of the soil. It

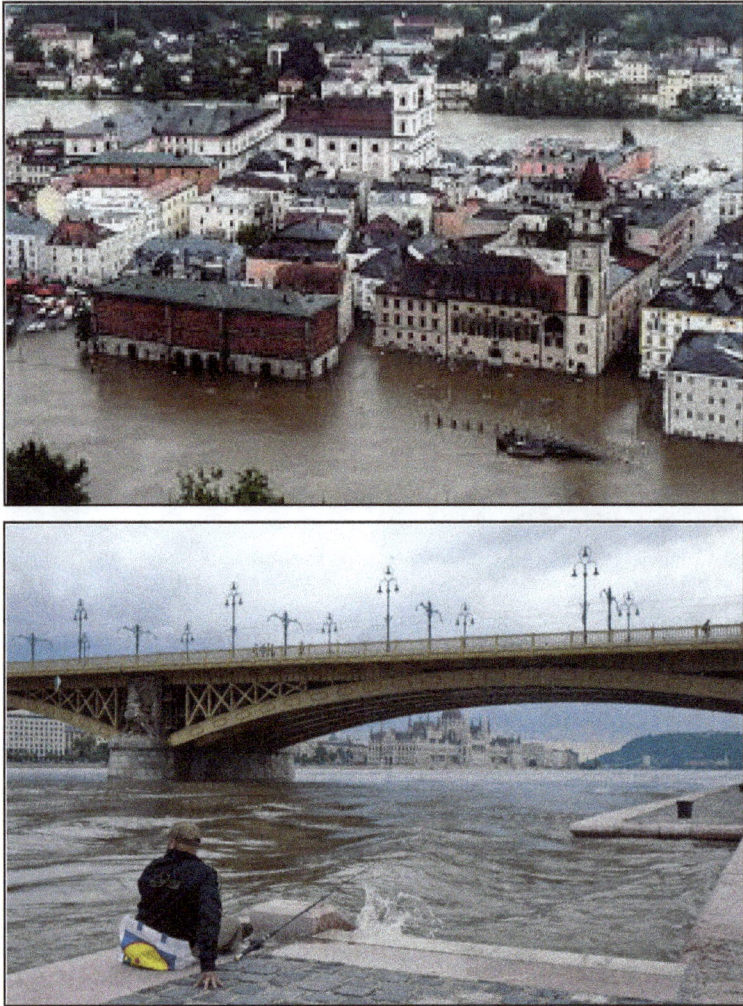

Figure 1.6: Flood in Central Europe in May, 2010.

is important that the actors of flood and excess water control and defence shall have appropriate plans, and possess the financial and technical means needed for action and maintenance of the respective works. It is also important to provide appropriate monetary resources for the investments needed for development of control and defence systems, bearing in mind the principle of tolerable damages (HAS, 2010).

References

1.   HAS, 2010: Climate change and Hungary: mitigating the hazard and preparing for the impacts - The VAHAVA report. Hungarian Academy of Sciences, Budapest, pp. 124.

2.   Móring A., 2011: Weather of 2010. *Légkör* **56**, 38-42.

3. Pongrácz R., Bartholy J. and Bartha E.B., 2013: Analysis of projected changes int he occurence of heat waves in Hungary. *Adv. Geosci.* **35**, 115-122.

4. Pongrácz R., Bartholy J. and Kis A., 2014: Estimation of future precipitation conditions for Hungary with special focus on dry periods. *Időjárás* **118**, 305-322.

URL

1. HMS, 2016: www.met.hu (18 February, 2016). Homepage of the Hungarian Meteorological Service.

Chapter 2

Global Warming and Climate Change Due to Poor Solid Waste Management: A Growing Concern in Indian Metropolitans

B.C. Prabhakar[1] and K.N. Radhika[2]

[1]Department of Geology,
Bangalore University, Bengaluru – 560 056, India
E-mail: bcprabhakar@rediffmail.com
[2]Department of Civil Engineering,
East West Institute of Technology,
Off Magadi Road, Bengaluru – 560 092, India

ABSTARCT

With the open door policies and free trade conditions prevailing, and acclaimed as one of the fastest growing economies, India is steadily drifting towards urbanization, and the present ~38 per cent urban enclaves of India are likely to expand to over 50 per cent in 2050, signaling a major leap awaited in the economic and industrial sectors. However, due to this urbanization, today, municipal solid waste (MSW) is posing a big threat in metropolitans in India, due to its colossal generation, but very poor management. With no effective legislation and mechanism to recycle the waste in sight, the problem is manifold increasing and haunting the municipal authorities and governments. An estimated 30 million tonnes of MSW is generated annually in Indian metropolitans. Unable to dispose the waste properly and in timely fashion, the heaps of waste are burnt callously, sometimes by public and sometimes by municipal authorities themselves causing severe air pollution. Abundant methane and carbon dioxide generated in MSW landfills are a potential cause for global warming, besides causing damage to ozone layer. But, if tapped, it (methane) can be a sustainable source of energy (biogas). Methane, resulting due to anthropogenic activity, especially from MSW is known to account for 10-12 per cent global warming. Thus it is very vital to realize that urban-induced effects are a significant contributors for climate change and the consequent global warming and the role of methane generated from the decay of garbage/waste is to be understood fully for its harnessing and proper management of MSW. Thus, the main objective of this paper is to review the alarming situation caused by improper management of MSW and to focus on the dire necessity of policy frameworks to handle this issue.

To ascertain the ground realities, the problems of MSW especially the garbage landfills, their adverse impact on soil, groundwater and air besides communities have been studied cursorily in Bangalore metropolitan, through field work, sample collection, and feedback from affected people.

Thus, in this paper, the authors are addressing an issue which is hitherto given not so much attention from the point of greenhouse gas emissions from the landfills of MSW and their steady contribution to global warming and in turn to climate change.

Keywords: *Global warming, Climate change, Municipal solid waste, Greenhouse gases, Methane emission, Urban policies, Indian metropolitans.*

Introduction

The planet earth, since its origin, has been evolved through the phenomenon of climate change. Climate change is also the driving force of evolution of life on earth. During the saga of this evolution, innumerable species have flourished and also perished, with long-period climate changes playing a critical role. Though climate changes and the consequent weather impacts causing cold periods (ice ages), warm periods, drought conditions *etc.*, are largely influenced by natural phenomena and operate in large cyclic periods, contribution of humans to climate changes is also being realized, though relatively on a low scale. Rapid increase in global population and paradigm shift in agrarian to industrialization have put enormous pressure on the resources especially energy resources, the consumption of which ultimately produces harmful additions to atmosphere. Today, human activity appears to be one of the significant sources for climate change, especially through his greenhouse gas loadings on the environment. As a consequence, the world is at an inflection point in the climate crisis. Recent anthropogenic emissions of green-house gases are the highest in history (IPCC, 2014). Due to the increased concentration of greenhouse gases in the atmosphere *i.e.*, CO_2 by 29 per cent, CH_4 by 150 per cent and N_2O by 15 per cent in the last 100 years, the mean surface temperature has risen by 0.4 – 0.8°C globally (IPCC, 2001). The consequences of climate change have been found to be severe in its several manifestations. The most telling impact has been spatial variability of precipitation *i.e.*, the increased fluctuation in intensity and frequency of precipitation. The other impact has been the sea level raise. It is recorded that the sea level has risen at an average annual rate of 102 mm in the past 10 decades (Sharma *et al.*, 2006). Based on the global assessment data since 1970, IPCC shows that anthropogenic global warming (60-90 per cent probability of occurrence) would have a discernible influence on many physical and biological systems (Rosenzwing *et al.*, 2008). Further, the report also notes that the global mean temperature may increase anywhere between 1.4 and 5.8 Celsius by 2100 (Panda, 2009).This change could transform the ecosystems with catastrophic disruptions on livelihoods, economic activity, living conditions and human health. Thus, research on the anthropogenic contributions to climate change has been one of the intensely pursued topic covering wide ranging human activities. However, the human contribution to climate change and global warming due to MSW is not addressed seriously.

The ever swelling population and ever increasing consumerist attitude world over, is ultimately responsible for generating colossal amounts of MSW which, through its ramifications, is silently yet steadily affecting the environment and ultimately contributing to global warming. Developing countries are the main contributors to this problem due to their poor management strategies of MSW. India being one such country, is facing the biggest problem of MSW management, especially of MSW generated in cities, and is bearing the brunt of this. In this paper, the authors have focused on the MSW management issues in the country, to highlight on the seriousness of the issue, and to strategize the management plans in order to prevent its adverse impacts.

Population, Urbanization and Waste Generation: Global Scenario

The global population at present is 7.4 billion (WPD, 2016). The developed countries with around 22 per cent of the global population produce more solid waste as compared to the developing nations, as they consume more than 60 per cent of the industrial raw materials (Elaine *et al.*, 2004).Developing countries, though with lower levels of per capita waste generated, the contribution of GHGs is higher due to the higher percentage (around 50 per cent) of organic (biodegradable) content. A report from the World Bank's Urban Development Department estimates that about 1.2 billion tonnes of waste is generated per year and this is likely to rise to 2.2 billion tonnes per year by 2025 with major contribution coming from developing countries (World Bank, 2012). The biodegradable waste generated in rural areas of the developing countries like India is reused as cattle feed or as fuel. But in the urban areas, the waste is either heaped up or landfilled, without following stringent scientific methods of waste disposal. A World Bank report on the global review of solid waste management shows that there were 2.9 billion urban residents in the world a decade ago which would swell to 4.3 billion by 2025 (Tahir *et al.*, 2015). They also show that the per capita municipal waste generated would also increase from 0.64kg to 1.42kg/capita/day.

Global Initiatives towards Effective Management of Municipal Solid Waste

The waste (MSW) sector is a significant contributor to greenhouse gas (GHG) emissions accountable for approximately 5 per cent of the global greenhouse budget (IPCC, 2006). Many advanced countries have achieved significant strides in managing the municipal wastes. Countries like Germany, Holland, USA, Canada and Australia have their own municipal waste management techniques wherein they "reduce, reuse and recycle" the waste. The Dutch's approach known as "Lansink's Ladder" is a very simple approach: avoid creating waste as much as possible, recover the valuable raw materials from it, generate energy by incinerating residual waste, and only then dump the left over in landfills but in an environmentally friendly way.

Globally, the composition of an urban Municipal solid waste would normally contain 51 per cent organics, 17.5 per cent recyclables like paper, plastic, metal, and glass and 31 per cent inerts. Its moisture content is 47 per cent and the average calorific value is 7.3 MJ/kg (1745 kcal/kg). When this waste is landfilled, the organic fraction in the waste slowly decomposes. In this process, landfill gas is formed which consists of 60 per cent methane (CH_4) and 40 per cent carbon dioxide (CO_2) with other trace gases.As landfills function as relatively inefficient anaerobic digesters, they provide significant long-term carbon storage, which is addressed in the 2006 IPCC Guidelines for National Greenhouse Gas Inventories (IPCC, 2006). GHG generation can be largely avoided through controlled aerobic composting and thermal processes such as incineration for waste-to-energy. Smith *et al.* (2001) have comprehensively quantified the benefits of recycling for indirect reduction of GHG emissions and report that the source-segregation of various waste components from MSW, followed by recycling or composting offers the lowest rate of greenhouse gas emission under assumed baseline conditions.Thus, when waste is handled scientifically and judiciously it will turn out in to a 'win-win situation'.

Population Explosion and Urbanization in India

India being the second most populous country in the world, is all set for rapid urbanization. The Census of India (2011) estimated a population of 1.21 billion which is 17.66 per cent of the world population. The population reference bureau (the global nodal agency studying the population growth rate) has projected that India will surpass China's population by reaching 1.6 billion by 2050. An estimate shows that by 2050, nearly 900 million people will be living in urban areas in India (Pathak, 2013). The level of urbanization of the country has increased from 17.6 per cent to 38 per cent in the last 50 years (Talyan *et al.*, 2008). Many researchers have also noted a phenomenal increase in population in Class 1 metropolitan cities (population of more than 1 million, as per 2001 census) and the number of metropolitans have increased from 23 to 35 in the last ten years (Figure 2.1).The economic growth of India during the last two and a half decades has been among the most rapid in the world. The last ten years have seen 6 per cent gross domestic product (GDP) growth per annum, and measured in purchasing power parity (PPP), India represents the fourth largest economy in the world today. The economic progress shows no sign of slowing down and is leading to rapid industrialization and population density in Indian cities.

The Problem of Municipal Solid Waste in Metropolitans in India

Generally, greater the economic prosperity and higher the percentage of urban population, greater is the amount of solid waste produced (Figure 2.2).The urban centres today, in India are generating around 50 million tonnes of MSW per year with a per capita amount of 125 kg per year, and a World Bank report (2012) projects an increase in waste to 100 million tonnes per year with a per capita amount of 255 kg by 2025. The class I cities like Delhi, Mumbai, Kolkata *etc.*, generate about 7000 tonnes of waste per day (Tahir *et al.*, 2015; PEARL, 2015) and Class II cities (population of between 50,000 to 99,999) is around 5000 tonnes. The municipal bodies in India are struggling to manage the waste due to limited waste recovery and processing

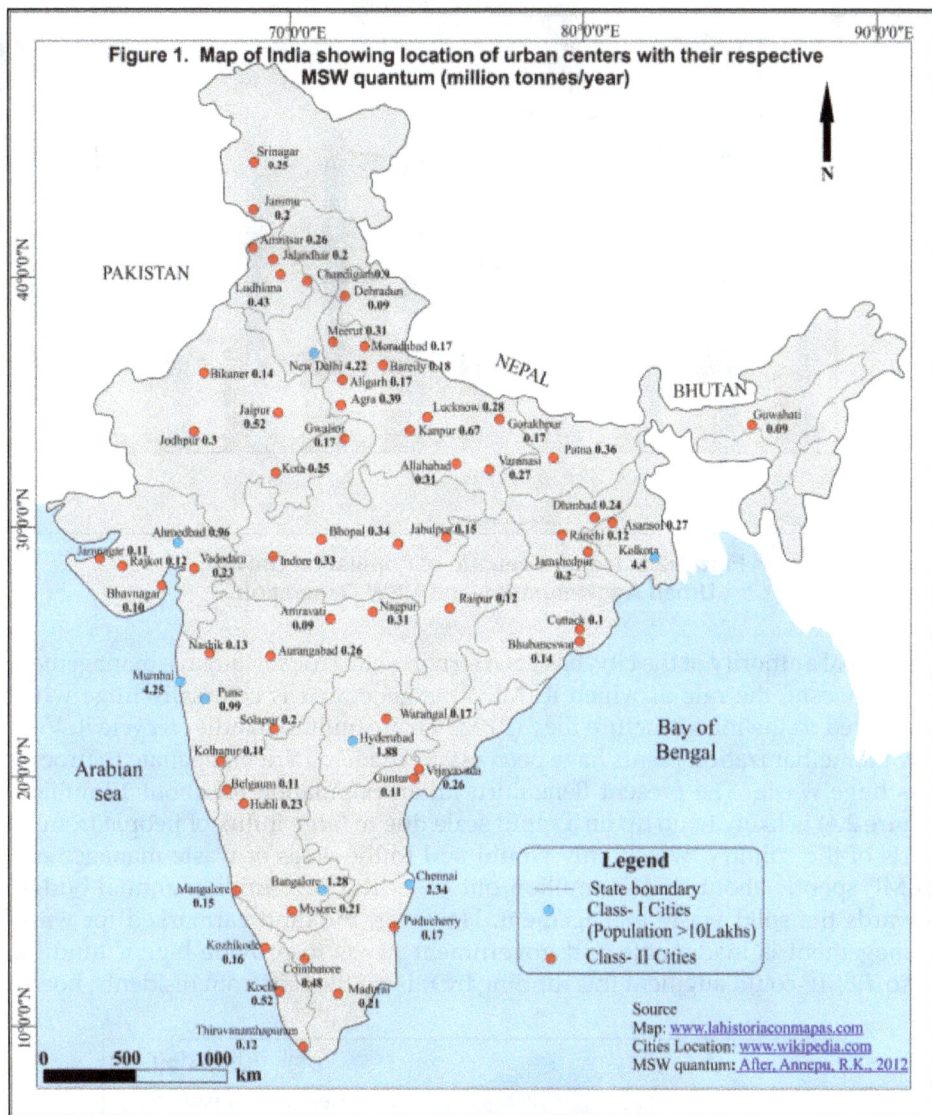

Figure 1. Map of India showing location of urban centers with their respective MSW quantum (million tonnes/year)

Figure 2.1: Map of India Showing Location of Urban Centers with their Respective MSQ Quantum (million tonnes/Year).

units and the waste is often dumped indiscriminately at open dump sites without leachate treatment (PEARL, 2015).

A preliminary study by Radhika and Prabhakar (2016) in the Bengaluru metropolitan area showed that MSW generated in Bengaluru is about 1.385 million tonnes per year out of which nearly 35 per cent of it has either recyclable or recoverable value. It's organic content is about 63 per cent (Figure 2.3) and when it is land-filled, it has high potential to generate methane. Though BBMP, the

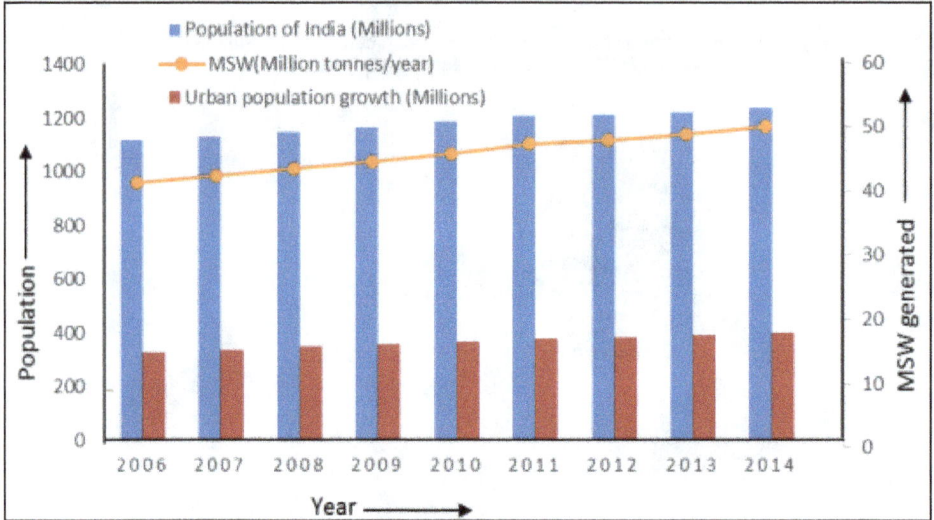

Figure 2.2: Indian Scenario of Population Growth, Urban Agglomeration and MSW Generation.

Municipal authority of the city, has been striving hard for sustainable management of this waste, the rate at which it (MSW) is generated is colossally huge when compared to the infrastructure they (BBMP) have built to handle/recycle it. Very few biomethanization plants have been established and are inadequate to process this huge waste. The present Bengaluru urban population (of about 10 million; Figure 2.4) is likely to go up on a rapid scale due to large influx of people from all parts of the country, which only would add to the woes of waste management. BBMP spends about Rs.4150 million out of its Rs.67290 million annual budget towards the solid waste management. However, the fund earmarked for waste management is inadequate and government needs to provide bigger funding. Also, BBMP could augment this funding by raising the cess from residents, hotels,

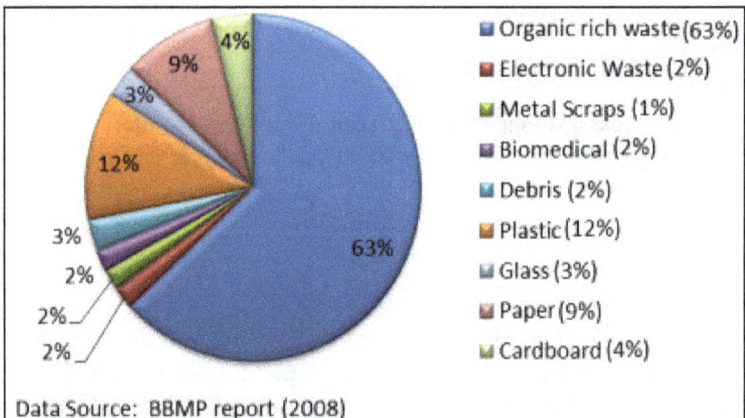

Figure 2.3: Composition of MSW of Bengaluru.

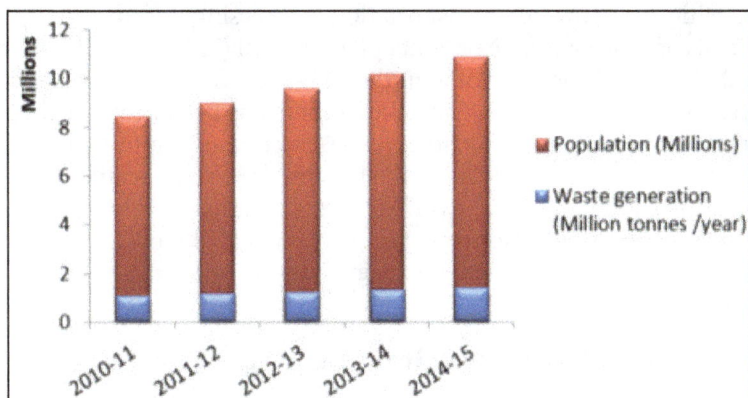

Figure 2.4: Population Growth and MSW Generated in Bengaluru.

marriage halls, industrial establishments *etc.*, to match the allocations required for efficient management of waste.

Further, an estimate based on the guidelines suggested by IPCC (2005) and the basic data from solid waste management master Plan (BBMP, 2008) suggest that Bengaluru's MSW can generate about 4274 tonnes of methane gas per year (considering the average population of Bengaluru to be around ten million and per capita waste generation to be around 0.35 kg/day). This means around 3,00,986 LPG cylinders, considering 14.2 kg gas in each cylinder, could be saved annually if the energy potential of the municipal waste is properly harnessed. The residue after harnessing the biogas could be used as organic manure.

The situation is no different in other major metropolitans of the country. The case of Mumbai struggling to manage MSW has come in to fore recently. Huge piles of MSW in Deonar dumping yard area of Mumbai were set on fire causing severe environmental pollution. The major Indian cities *i.e.* class 1 and 2 put together account for about 450 million population of the country and generate about 50 million tonnes of MSW, which in turn account for 73 per cent of country's MSW production. From the landfills of this MSW about 1,50,000 tonnes of methane (based on first order decay method, IPCC, 2005) could be generated. Importantly, harnessing this methane would reduce GHG emissions to atmosphere.

Policy and Legislative Framework for Municipal Solid Waste Management in India

The legislative framework for municipal solid waste management was initiated in the fourth 5-Year plan (1969–74). Later, in 1975, the Government of India appointed a high-level committee to review the problem of urban solid waste in India. This committee made 76 recommendations covering 8 important areas of waste management. However, the most significant piece of legislation is the Municipal solid waste (management and handling) rules, 2000. The major criteria in the rules was to segregate the waste into different types based on source and composition and further, transportation, processing and disposal should be planned in accordance with the national plan. Asnani (2004) carried out a study to ascertain the status of compliance with the MSW Rules 2000 by major cities as on 1.4.2004. The results of this study reveal that there has been insignificant progress in the matter of processing of waste and construction of sanitary landfills. Many cities have not even initiated the implementation of the rules even though the time frame prescribed in the rules was over. Major constraints for non-compliance were unavailability of financial resources, lack of technical skilled workforce, lack of public awareness and motivation, and non-cooperation of the households, trade and commerce.

India: The Impact Due to Climate Change

Due to global warming, the climate pattern of India is changing day by day with increase in climate disasters. A statistics says that 27 out of 35 States are disaster-prone, with floods being the most frequent disaster (Barbhuiya, 2015). According to the Indira Gandhi Institute of Development Research, continued global warming would result in climatic disasters, which in turn would cause a decrease in India's GDP to decline by about 9 per cent, with a decrease by 40 per cent of the production of major crops. A temperature increase of 2°C in India is projected to displace seven million people, with a submersion of the major cities like Mumbai and Chennai. According to surveys, India ranked third highest with 18 disasters in 2007-08, resulting in the death of 1103 people (Rajasekhar, 2012).Thus climate change is likely to have negative impact in the form of unpredictable rainfall pattern, melting of glaciers, droughts and expanding seas. One of the major challenges to coup this kind of situation is to ensure food security, besides handling hazards. As one of the mitigating measures to reduce global warming and climate change, the MSW problem has to be addressed seriously.

Poilcy and Legislative Framework for Climate Change in India

Anthropogenic climate change poses perhaps the most complex policy issue faced yet by the global community. At present, India is contributing about 5.7 per cent of the global carbon emissions. Most carbon emissions arise from use of coal in electric power and industry sectors, while agriculture and livestock sector emit substantial amounts of methane. India has already witnessed the alarming effects of its carbon emissions which would add to the woes of global warming. Carbon mitigation in India is complicated by the fact that India has large coal reserves and it continues to be the chief fuel mineral for energy resource. The substitution away from coal therefore would require energy imports. While India has the experience with emerging renewable energy technologies, the capital and foreign exchange constraints are likely to restrict the shift away from coal, unless the economic and fiscal policies to relax these constraints. With this coal consumption is expected to continue for some time, India, inevitably, is unlikely to reduce its GHG emissions to atmosphere. This woe is compounded by methane and other harmful gas emissions from MSW landfill sites which puts India in a tight spot in a global scenario. Hence, policies covering holistic approach is the need of the hour to reduce the India's contribution to climate change. It obviously means that all sources of GHG emissions have to be prioritized to manage them stringently.

Discussion and Conclusions

With global warming and its manifold adverse effects, climate change has taken the centre stage of present day research and discussion at global level. Though the developed countries are the main contributors to this situation, it is the developing countries on whom pressure lies to take stock of the situation and to take mitigating measures. The key issue is that this has to be achieved without sacrificing the developmental programs (which are mainly energy dependent, and carbon based) and at the same time reducing the emission of GHGs. Besides several strictures already in vogue for variety of industrial and vehicular-related emissions, both technology-wise and legislation-wise, there is an urgent need to draw comprehensive policies and long-term implementation strategies to tackle the issue of MSW on a sustainable basis, which otherwise would not only continue to pose a nuisance value, but also contribute for the global GHG emissions.

With about 1 per cent raise in urbanization in India and about 1.5 per cent raise in MSW annually in Indian metropolitans, MSW-related GHGs are sure to add to climate woes. Studies in some of the landfill areas of Bengaluru metropolitan have clearly shown their potential for methane which can be tapped as biogas, a source of energy. If untapped, this methane has 25 per cent more capacity than CO_2 in causing global warming besides unleashing other hazardous gases. The preliminary studies in this regard on Bengaluru are also perceptible for ever growing urban enclaves in India. Unfortunately, the civil laws and the implementation machinery in the country are at a total disarray in the management of MSW. The short sighted actions in handling it are no more helping the situation.

The most urgent action required is to restrict the unplanned growth of cities. The urban planning has to be rationalized through policies to lay stress on developing newer and smaller cities with sustained resource base. The over sized metropolitans like that of Bengaluru, Mumbai, Chennai and others have to be prevented from further growth especially the mushrooming residential lay outs (they are the real bane of urban growth), who have menacingly contributing for the MSW. One of the ways to check them is to prevent the establishment of new industries, new companies, new government offices and so on. Spreading out such establishments to newer or existing smaller cities will not only help in better handling of MSW, but also contribute for the growth of economy in less developed areas. Restricting unplanned growth of cities also assumes importance from the point of minimizing the impacts (on life and property), in the event of a natural disaster. The Bengaluru city never had a problem of flooding till 1960s, but due to its rampant and chaotic growth in all directions, the drainage streams, tanks *etc.*, have been choked (largely due to MSW littering and dumping), and the residents dread floods during heavy down pours. The unsustainable growth in this city has also drained out the precious water resources. So, checking the growth of oversized cities has become an immensely important issue for minimizing the impacts of global warming and climate change. Time has also come to make radical changes in urban policies so that substantial powers are rested with Government authorities so that they can quickly and effectively overcome the obstacles in implementing public amenity programs. Because the present policies are beset with too many provisions which extend a great deal of freedom (through laws) for individuals to hamper the process of implementation and management of issues concerning common public.

Thus the MSW has to be prioritized as a national issue for its proper management, which is direly necessary from the point of ensuring cleaner environment, cleaner living, public health, aesthetic value, harnessing energy and above all preventing global warming.

Abbreviations

MSW: Municipal Solid Waste

IPCC: Intergovernamental panel of Climate Change

BBMP: Bengaluru Bruhat Mahanagara Palike

SWM: Solid Waste Management

GHG: Greenhouse gases

References

1. Annepu, R.K., 2012. Sustainable solid waste management in India. Thesis submitted to Columbia University, p. 189, Weblink: www.swmindia.blogspot.in

2. Asnani, P.U., 2004. United States Asia Environmental Partnership Report, United States Agency for International Development, Centre for Environmental Planning and Technology, Ahmedabad.

3. Barbhuiya, F., 2015. Natural Disaster, Especially Flood and Its Management in India: With Special Reference to Assam. International Journal of Humanities and Social Science Studies. Volume-II, Issue-III, pp. 289-299. 2349-6959 (Online), ISSN: 2349-6711 (Print).

4. BBMP, (Bruhat Bengaluru MahanagaraPalike), 2008. Solid Waste Management in BBMP. Weblink: http://218.248.45.169/download/health/swm.pdf

5. Census of India, 2011. Figures at a glance, India – provisional. pp.179. Weblink: www.scribd.com

6. Elaine, B., Bournay, E., Harayama, A., Rakecewicz, P., Catelin, M., Dawe, N., Simonett, O., 2004. "Vital Waste Graphics." United Nations Environment Program and Grid-Arendal. p. 48. < http://www.grida.no/publications/vg/waste/page/2853.aspx

7. IPCC (Intergovernmental Panel on Climate Change), 2001. Climate Change 2001: The Scientific Basis, Contribution of Working Group I to the Third Assessment Report of the intergovernmental Panel on Climate Change (IPCC), Cambridge University Press, Cambridge.

8. IPCC (Intergovernmental Panel on Climate Change), 2006. IPCC Guidelines for National Greenhouse Gas Inventories. www.ipcc-nggip.iges.or.jp/public/2006gl/

9. IPCC (Intergovernmental Panel on Climate Change), 2014. Climate Change 2014 Synthesis Report Summary for Policymakers, p. 31. Source: https://www.ipcc.ch

10. IPCC, (Intergovernmental Panel on Climate Change), 2005. Calculation Tools for Estimating Greenhouse Gas Emissions from Wood Product Facilities. National Council for Air and Stream Improvement, Inc. Weblink: www.ghgprotocol.org

11. Panda, A., 2009. Vulnerability to climate change and planned adaptation: Data requirements and Gaps in India. Paper published in "Proceedings of the national seminar on Climate change, Data requirement and availability. Organized by Institute for social and economic change, Bangalore" Edited by, Nautiyal, S., and Nayak, B.P. pp. 115-130.

12. Pathak, H.C., 2013. Urban and Peri-urban agriculture, Policy paper 67, National academy of agricultural sciences. Pp.12. Weblink: http://www.naasindia.org/

13. PEARL (Peer Experience and Reflective Learning), 2015. Urban Solid Waste Management in Indian cities. A report submitted to National Institute of Urban Affairs by Urban Management Consulting Pvt. Ltd. and Center of Environment Education. Pp.96. Weblink: http://www.pearl.niua.org

14. Radhika, K.N. and Prabhakar, B.C., 2016. Urban enclaves and their contribution to global warming through municipal solid waste – innovative approaches in Bangalore Metropolitan. Proceedings of KSTA 2016 conference on Energy, Climate change and Environment held on 29th and 30th of January 2016. pp.11-12.

15. Rajasekhar., K., 2012. Climate change, natural disasters, mitigation and preparedness in Indian continent: A study. Asian Journal of research in Social Sciences and Humanities. Volume 2, Issue 11, pp.32-42. ISSN:2249-7315 (online).

16. Rosenzwing, C., Karoly, D., Vicarelli, M., Neofotis, P., Wu, Q., Casassa, G., Menae, A., Root, T.L., Estrella, N., Seguin, B., Tryjonowski, P., Chunzhen, L., Rawlins, S., and Imeson, A., 2008. Attributing physical and biological impacts to anthropogenic climate change. Nature 453, pp. 353-357.

17. Sharma, S., Bhattacharya, S., and Garg, A., 2006. Greenhouse gas emissions from India: A perspective. Current Science, Vol.90, No.3, pp.326-333.

18. Smith, A., Brown, K., Ogilvie, S., Rushton, K., and Bates, J. 2001. Waste Management Options and Climate Change. Final Report ED21158R4.1 to the European Commission, DG Environment, AEA Technology, Oxfordshire.

19. Tahir, M., Hussain, T., Behaylu, A., 2015. Scenario of Present and Future of Solid Waste Generation in India: A Case Study of Delhi Mega City. Vol.5, No. 8, pp. 83-91.

20. Talyan, V., Dahiya, R.P., Sreekrishnan, T.R., 2008. State of municipal solid waste management in Delhi, the capital of India. Waste Management. pp. 1276–1287.

21. World Bank, 2012. What a waste: a global review of solid waste management. Urban development series knowledge papers. http://siteresources.worldbank.org/INTURBANDEVELOPMENT/Resources/336387-1334852610766/What_a_Waste2012_Final.pdf

22. WPD (World population datasheet), 2016. World population datasheet report, Washington DC, United States of America, pp.22.Weblink: www.worldpopdata.org

Chapter 3

The Impact of Climate Change on Natural Disasters in Iran

Mahdi Faravani[1], Reza Aghnoum[1], Ali Akbar Moayedi[1]
and Muhammad Ehsan Elahi[2]

[1]Faculty Member,
Razavi Khorasan Agricultural and Natural Research and Education Center, Iran
E-mail: mfaravani@gmail.com
[2]Scientific Officer,
Arid Zone Research Centre, Pakistan Agriculture Council, Pakistan
E-mail: niazidk@hotmail.com

ABSTRACT

In Iran, the largest source of greenhouse gas emissions (82.23 per cent) is from energy sector, burning fossil fuels for electricity, heat and transportation. Other most important emitter is commercial and residential, land use change, industries and agriculture. Emissions of greenhouse gases are already changing the climate of Iran. The consequences of climate change have caused numerous problems for different sectors in recent years. It has resulted in severe droughts on one hand and on the other disasters such as floods, strong storms and hurricanes are frequent common these days. Over the last decade, people have witnessed at least eight years of drought in Iran and neighboring countries. The drought has severe implications, particularly in important cereals production. One problem that people shall face will be the higher price of cereals. Strong storms, hurricanes, floods, and rising sea levels in coastal areas are the other serious concerns. Therefore, it is necessary to identify/study negative impacts of this phenomenon and solutions for their adaptation and mitigation should be considered in scientific and administrative plans of the country. This paper discusses the overall impact of climate change on various types of natural disasters in Iran.

Keywords: *Floods, Storms, Mitigation, GHG, Drought, Hurricane.*

Introduction

Countries that industrialized generations ago like the U.S. and members of the European Union now face expensive job of repairing and upgrading their aging critical infrastructures such as from roads and bridges to energy and water systems. Bridges are in danger of falling, water pipes are leaking, and electric grids are vulnerable to a wide range of risks including cyber-attack, extreme weather and inadequate maintenance.

There is an increasing concern worldwide about the ineffectiveness of current drought management practices that are largely based on crisis management. These practices are reactive and, therefore, only treat the symptoms (impacts) of drought rather than the underlying causes for the vulnerabilities associated with impacts (Wilhite *et al.*, 2014).

Climate change has extensive impacts on the Earth's climate system in the previous periods. Major impacts are: the impact on the atmosphere, hydrosphere, ice sphere and the impact on the biosphere. Human activities will increase the amount of emission of greenhouse gases in the coming years and decades and subsequently future temperatures are expected to increase to 1.5°C to 4.5°C depending on the global emission rate by the year of 2100 (Figure 3.1) (O'Neill and Oppenheimer, 2002). Thus, with more emit, the future major disasters such as acidic ocean, higher sea levels, and larger changes in precipitation and drought patterns will happen with continuing emissions of greenhouse gases that will lead to a warmer atmosphere and also changes in earth's temperature ultimately impacting the global carbon cycle. The extent of future climate change depends on what we do now to reduce greenhouse gas emissions (Smith *et al.*, 2016). Climate change is likely to influence on tourism industries and the locations preferred by tourists in Iran during 1961-

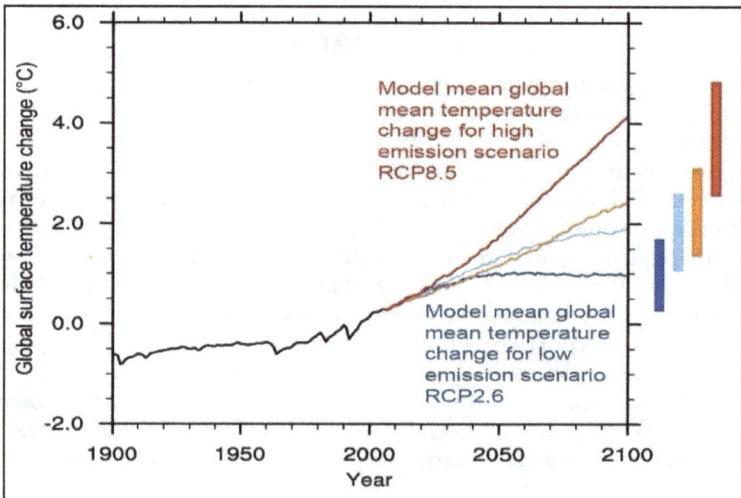

Figure 3.1: Changes in Projected Global Average Temperature under Emissions Pathways Model for the Observed Data from 1986-2005. (*Source*: IPCC, 2013).

2010. This predict can be useful to identify the suitable locations for tourism, now and in the future, to guide strategic investment (Roshan *et al.*, 2016).

The Islamic Republic of Iran is rich in energy resources and oil has dominated the politico-socio-economic life of this country, but nevertheless Iran had no documented policy and vision for its massive resources. This poses a major challenge to policy makers and the society in facing with the impact of climate changes (Tofigh and Abedian, 2016). Iran has shown remarkable growth in total fossil-fuel CO_2 emissions since 1954, averaging 6.3 per cent per year. Total emissions has reached to 147 million metric tons of carbon in 2008. With Iran being the world's fourth largest oil-producing country it is not surprising that crude oil and petroleum products account for the largest fraction of the Iranian emissions, 46.4 per cent in 2008. Emissions from gas fuels have grown 390-fold since 1955, and now account for 42.3 per cent of Iran's total fossil-fuel CO_2 emissions which is above the global average at 2.00 metric tons of carbon from a per capita standpoint (Figure 3.2). This CO_2 cause air pollution. Recently, air pollution is a problem affecting Iran's major and over crowded cities. Pollution free holidays are very uncommon nowadays. According to Iran's deputy health minister around 4,500 people died due to air pollution in the first nine months of 2013. As of 2008, about 77 per cent was used for transportation and by household and commercial purposes, 6 per cent by industries, forest 2 per cent, waste 6 per cent, and 9 per cent by agriculture and other activities (Figures 3.3 and 3.4) (Shahbazi *et al.*, 2016).

According to COP21,the main aim is to achieve a legally binding and universal agreement on climate, for keeping global warming below 2°C. The agreement is due to enter into force in 2020. For this goal, the new technologies are needed to remove greenhouse gases from the atmosphere. By 2050, the average temperature of Iran is projected to increase by about 1.5°C, depending on emissions scenario

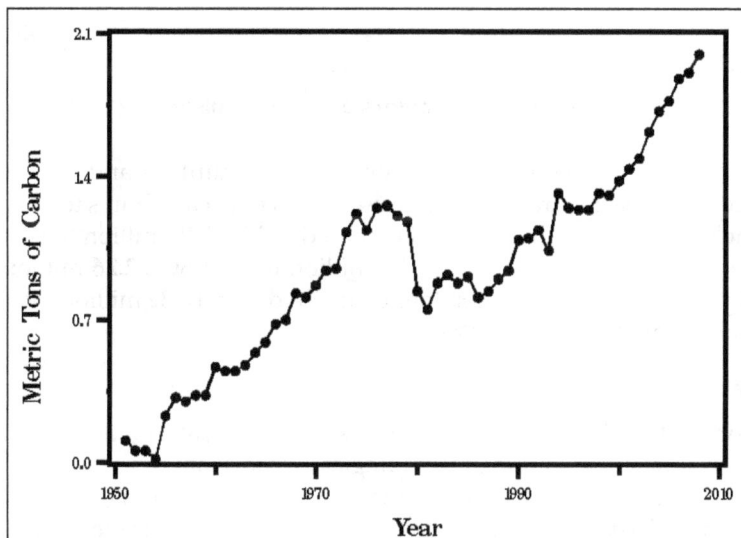

Figure 3.2: An Estimate for per Capita CO_2 Emission in Iran.

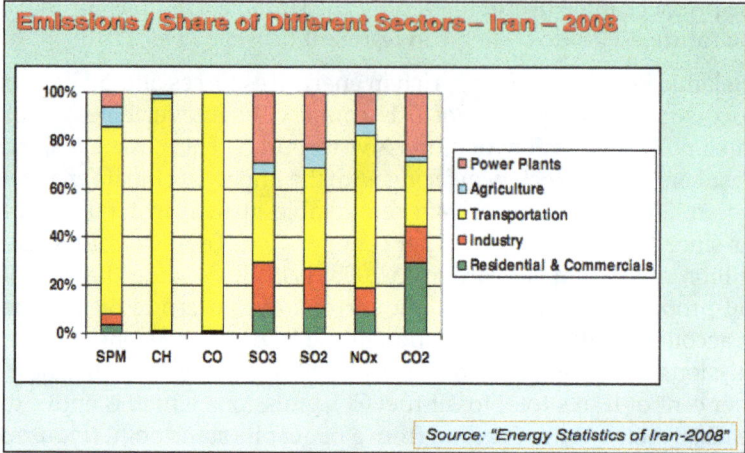

Figure 3.3: The Diagram above Demonstrates the Majority of CO$_2$ Emissions can be Attributed to Residential and Commercial Sectors as well as to the Transportation Sector.

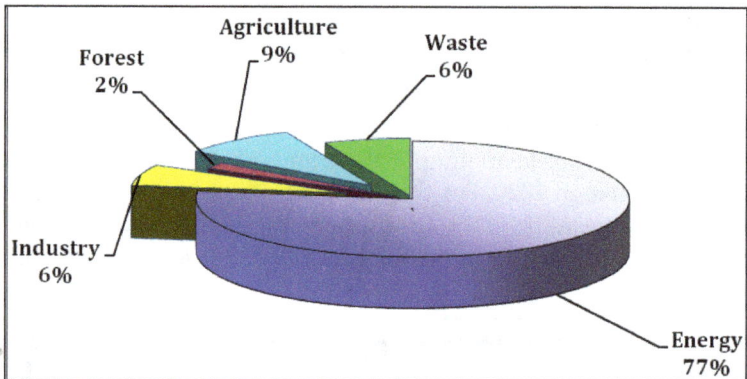

Figure 3.4: Contribution of different Sectors on GHGs Emissions (1994 - CO$_2$ Equiv.).

and climate model. A concern about energy consumption and consequential environmental impact in Iran has been raised in recent years. Iran's total emissions in 2006 includes 413.23 million tons of carbon dioxide, 2.18 million tons of carbon monoxide, 2.5 million tons of NOx, 0.75 million tons of SO$_2$, 2.26 million tons of residual hydrocarbons, 0.59 million tons of aldehydes and 0.32 million tons of dusts (Avami and Farahmandpour, 2008).

Materials and Methods

The present study is to review the work and formulate a massive action plan regarding adaptation and mitigation strategies in Iran. It analyzes the research gaps to manage drought risk in changing climate scenario. In this case, there is focus on disasters from 50 years ago on the development plans and implement programs leading to reduction of GHGs and to manage the adverse impacts of climate change over water resources, agriculture and forestry, human health, biodiversity and coastal zones in Iran and future implications of natural disasters. It includes data collection from the observations and interview with the district civil administration and local residents.

Results and Discussion

Except under the most aggressive mitigation scenario studied, global average temperature is expected to warm at least twice as much in the next 100 years as compared to the past 100 years. Climate models predict the following key temperature-related changes in the atmosphere temperature. Its increment will intensify changes in the Earth's climate variables. Even if emissions of entire greenhouse gas are stopped at once, but due to their already released amounts into the atmosphere, the changes in the Earth's climate variables will be evident. The negative consequences of changing in climatic indices could be more destructive for human beings than the 10 factors threatening mankind in the 21st century such as poverty, nuclear weapons, food, *etc.*

In Iran, the consequences of climate change have caused numerous problems for different sectors in recent years. The most important are increased atmospheric-climatic disasters like droughts, floods, dust storms, heat waves, *etc.* Therefore, it is necessary to better identify and study negative impacts of this phenomenon and mitigation solutions should be considered in scientific and administrative policy of the country.

National Circumstances of Iran

Iran is the second largest economy in the Middle East and North Africa (MENA) region after Saudi Arabia, with an estimated Gross Domestic Product (GDP) in 2014 of US$406.3 billion. It also has the second largest population of the region after Egypt, with an estimated 78.5 million people in 2014 while in 1994 was about 57.7 million. It is a core factor of risk generation. They imply the settlement of people and assets that can lead to an increase of the exposure and the vulnerability to natural hazards. The urban population (72.32 per cent), rural population (27.680 per cent), population density (47.6 People/km^2) and urban population growth (2.094 per cent annual) are important index to consider as often accompanying unmanageable sprawling and unmanaged settlements, can be seen as a driver of disaster risk, since rapid and unregulated urban development can contribute to the concentration of people and assets in hazardous locations (Tables 3.1–3.3).

Poverty Conditions

In 2005, poverty was 1.45 per cent in Iran using a poverty line of US$1.25 per day (PPP). World Bank projections estimate that only 0.7 per cent of the population (half a million people) lived under this poverty line in 2010, although a large proportion of people are living close to it. Indeed, raising the poverty line by US$0.5 (from US$2 to US$2.50 and from US$3 to US$3.50) could put 4 per cent -6 per cent of the population (over 4.5 million people)under poverty. This suggests that many individuals are vulnerable to changes in their personal disposable income and to the persistent rise in the cost of living.

Table 3.1: Basic Iran Population Statistics and Indicators

Population	million people	77,447,168
Urban	per cent Total population	72.320
Rural	per cent Total population	27.680
Urban population growth	per cent Annual	2.094
Population density	People/km²	47.6

Table 3.2: Reported Losses of Iran and International Mortality during 1990–2014

	Per cent		
	Earthquake	*Floods*	*Other*
Nationally mortality	92.1	6.5	1.5
Internationally mortality per cent	97	2.8	0.2

All scale disasters without criteria.

Table 3.3: Iran and Internationally Reported Economic Losses during 1990 - 2014 EMDAT

	Per cent			
	Drought	*Earthquake*	*Floods*	*Other*
Nationally combined economic losses	45.5	30.4	23.7	0.4
Internationally reported losses	16.5	53.5	29.8	0.2

Source: EM-DAT (Feb. 2015) - The OFDA/CRED - International Disaster Database http://www.emdat. be – Université catholique de Louvain Brussels – Belgium

Iran's economy is characterized by a large hydrocarbon sector, small scale agriculture and services sectors, and a noticeable state presence in manufacturing and financial services. Iran ranks second in the world in natural gas reserves and fourth in proven crude oil reserves. Economic activity and government revenues still depend to a large extent on oil revenues and therefore remain volatile. In this regard, the agricultural research and education center along with other related organizations such as environmental department and metrological organization try to put the more precise understanding and awareness of modern methods in detection, assessment, adaptation and mitigating the negative consequences of climate change on the agenda while acquiring the latest scientific findings.

The area coverage of different types of climate in Iran is 35.5 per cent hyper-arid, 29.2 per cent arid, 20.1 per cent semi-arid, 5 per cent Mediterranean, and 10 per cent wet (of the cold mountainous type). Thus more than 82 per cent of Iran's territory is located in the arid and semi-arid zone of the world. The average rainfall in Iran is about 250 mm, which is less than 1/3 of the average rainfall in the world (860 mm). In addition, this sparse precipitation is also unfavorable with respect to time and location. Another important climatic element is extreme temperature changes that sometimes range from −20°C to +50°C. Severe drought is also recognized as a feature of Iran's climate. In the last three years, the country has suffered severe desiccation and this lack of rainfall has resulted in extensive losses.

Observation of the Climate Change Impacts on Iran

When analyzing the country's current temperature trend and in the forecasting's, specifically related to time, a significant temperature increase was observed during the summer months. Over the last decade, we have witnessed at least eight years of drought in Iran and neighboring countries. The drought has severe implications, particularly in grains and cereals production. One problem that people shall be faced with will be the price of cereals. Strong storms and hurricanes, and floods, and also the rising sea levels in coastal areas are other serious problems (Figure 3.5).

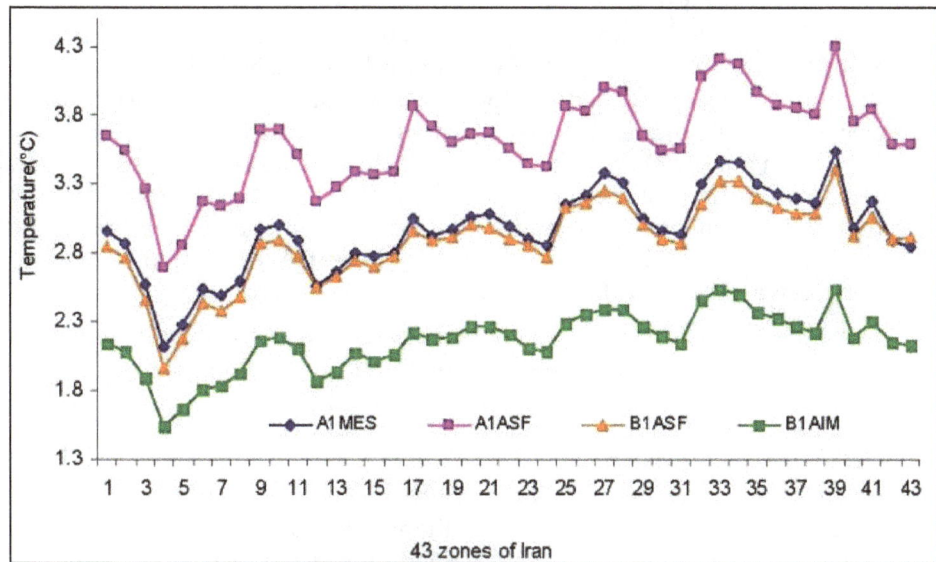

Figure 3.5: Total Average of Temperature Annual and Seasonal Changes from the Year 2025 to 2100 Based on the Results of different Scenarios for the Considered Regions.

Also, with regard to altitudinal levels, it was evident that stations at higher altitudes show a more significant increase in daily and mean daily temperatures. Taking into account the output mean of the different climate change scenarios, the temperature simulations show a 4.41° C increase in Iran's mean temperature by 2100. Most of these temperature increases would occur in the southern and eastern parts of Bushehr, certain coastal regions of the Persian Gulf, eastern and western parts of Fars, Kohgilooye, Boyerahmad, southern parts of Yazd, as well as southern and southeastern parts of Esfahan (Roshan *et al.*, 2011).

Iran is in a highly vulnerable environmental state right now, and the country is currently facing a severe water shortage. The most important issue to cut greenhouse gas is a revolution in the agriculture systems from the current practices, sowing dates, irrigation methods and crops variety, they need to be adopted and adjusted to this climate situation. More than 90 per cent of water is used for agriculture in Iran.70 per cent of the problems is due to the inefficient consumption of water in this sector. So the government is investing on new technology and providing incentives, awareness/education and training for the farmers that are the major strategy that can be used to mitigate the impact of climate change. There is increased frequency, severity and duration of drought events in association with a changing climate and as well other human activities and policy makers. This is showed some of the most important occurring disasters of nature in Iran.

Iran's Lake Orumieh, which lies in the northwestern Iran is the second largest salt water lake on earth, has been shrinking due to ecological factors and human activity. It is the largest inland body of water in Iran, with a surface area of approximately 5,200 square kilometers. Its deepest point is approximately 16 meters deep. It is home to some 212 species of birds, 41 reptiles, 7 amphibians, and 27 species of mammals, including the Iranian yellow deer. Experts have warned that the construction of a 1.5 kilometer bridge over the lake, finished a few months ago, together with a series of ecological factors, will eventually lead to the drying up of Orumieh Lake. Declining rainfall, climate change and climate warming accelerate the evaporation process of the Lake, which raises to global concerns over the issue. The disaster may turn the lake into a salt marsh in less than 20 years which will directly affect the climate of the region. The construction of a dam on part of the Lake and the recent drought has significantly decreased the annual amount of water Orumieh receives. The level of salt in Orumieh Lake has reached 400 grams per litre.

Blame the gradual drying of the salt lake on the government and its policies of damming rivers, but officials say drought and global warming has caused the disaster.

Zayandeh Rud's Tragedy

Zayandeh Rud is the most important catchment in the central Iran. This basin is located in arid and semi-arid region with the annual rainfall of 50 mm. It is a very important catchment which plays a vital role in the center of Iran. It is a crucial water resource for irrigation, industries, animal farming and municipal supply, agriculture is the most fundamental and conventional sector which exploits most of the Zayandeh Rud water resources. Studies revealed that the volume of the Zayandeh Rud dam that in general is more than 700 million cubic meters, currently it is about 190 million cubic meters which can show the critical condition of the area itself. The major reasons of water shortages are recent years 'drought, increase of consumption, decrease of resources, and multi-management of the basin. Since different sectors attempting to manage the catchment without any coordination and cooperation with others, the demands and needs are different and these worsen the situation. Overloading on Zayandeh Rud which is due to water transfer from Zayandeh Rud River to the other provinces.

Agriculture has typically been the first and most affected sector on the aforesaid disaster but many other sectors, including energy production, tourism and recreation, transportation, urban water supply, and the environment, have also experienced significant losses and decreasing water amount. The Iranian plateau is subject to most types of tectonic activity, including active folding, faulting and volcanic eruptions. It is well known for its long history of disastrous earthquake and as well the most seismically active countries in the world, being crossed by several major fault lines that cover at least 90 per cent of the country. Its long history of disastrous earthquake activity Iran is one.

Other disasters in Iran occur in some part of Iran are tropical cyclones (Cyclone Gonu and Cyclone Yemyin) and hurricanes. During the spring of 2014, heavy rain and hailstorms struck the Iranian capital Tehran, surprising residents and causing traffic jams across the city. The fierce June 2 hurricane packed with thunder and lightning, battered the northern parts of Tehran and lasted for more than an hour. According to Institute of Geophysics, wind speed was 80 km/h; Meteorological Organization of Iran reported 120 km/h. Air pressure was 4 mbar. The worst dust storm in the area was reported 2 weeks ago in Ahwaz with 10 thousand microgram per cubic meter of dust in air, 66 times the healthy limit.

Climate Change-Government Action

Carbon Sequestration to Mitigate Climate Change

Carbon sequestration in the decertified rangelands of Hossien Abad, south Khorasan, through community-based management. The main global objectives are to sequester C, to improve the ecosystem through natural regeneration by planting/ seeding drought resistant grasses and shrubs, and to make the rangeland areas of Iran more productive, bringing thereby most of the area under vegetation cover.

Yazd and Shiraz Solar Energy

The Yazd integrated solar combined cycle power station is a hybrid power station situated near Yazd, Iran which became operational in 2009, and in 2011 as a solar integrated plant. Shiraz solar power plant is a concentrating solar power type pilot power station situated near Shiraz, Iran.

Conclusions

Similar to many countries in the world that experiencing extreme water shortages, the Islamic Republic of Iran is also in the midst of a serious water crisis. The country's resources management is facing many subsequent challenges, including growing demand for water resources with proper quality, a considerable increase in the costs of supplying additional water and urgent need to control water pollution as well as the uncontrolled exploitation of underground waters and the necessity to conserve these valuable resources. If immediate mitigation measures are not taken, the situation could become even more disastrous in the years to come. Being cognizant of the crisis' importance and its destructive influences, governmental authorities have begun evaluating their plans and programs and have devised long-term strategies to allay the water crisis. Some important policies that the government has to develop like support transitioning from non-efficient technology to more sustainable practices through capacity building and knowledge transfer and development of climate change sector specific support rather than budget support to ensure funds are targeted in areas which needed.

This paper describes the state of the country's threatening water crisis while also explaining in detail its main sources and the subsequent challenges it has caused the country to endure. Discussions on how governmental authorities are fighting the crisis and the state of the approaches which they have pioneered are also presented. Moreover, context of the Long-term Development Strategies for Iran's Water Resources is discussed from sustainable water management point of view with practical recommendations on the issue.

Iran has a more advanced water management system than most Middle Eastern countries, similar to the other countries in the region but Iran is experiencing a serious water crisis. The looming crisis is being blamed on a number of factors including population growth and uneven distribution, natural phenomena such as droughts and changing climate patterns, and the mismanagement of existing water resources who believing that water shortages are periodic (Madani, 2014). Curbing greenhouse gas emissions is a prerequisite to reducing disaster risk efficiently, as climate change has already raised the number and intensity of many natural hazards. By using renewable energy, low carbon technologies, clean development mechanism (CDM) projects, aforestation, reforestation, water use efficiency, land use change policy, desertification, nuclear energy *etc.*, we will be able to play a more active role in curbing greenhouse gasses and improving the quality of air in the major cities.

There are lots of rules, regulations and plans on the subject of earthquake and the impact of the climate change on the disasters, but a few of them have been implemented in vulnerable area so far. There is increasing concern worldwide about the ineffectiveness of current drought management practices that are largely based on crisis management. Development of water savings irrigation techniques for promoting modern agriculture construction, conservation oriented society has a very important significance. Water savings irrigation techniques significantly more than the traditional irrigation techniques to save water. The combination of climate change and land use change bring changes in the amount, timing and intensity of rain events. On the other hand increasing population also increase the demand for water, water savings techniques help us to save water and introduce re-use of water.

There is an urgent need for Iranians to execute their water resources development strategies and risk management instead of crisis management, since prevention is always better than trying to cure. It is also necessary to consider the population growth, mitigation and population distribution as the main cause of this crisis. The time for adopting an approach that emphasizes drought risk reduction is now, given the spiraling impacts of droughts in an ever-increasing number of sectors and the current and projected trends for the increased frequency, severity and duration of drought events in association with a changing climate. Changes in disaster frequency can be the result of both an increase in actual occurrences of a hazard and an increase in human activity where the hazard already exists. Frequency of the international disaster, earthquake (47.2 per cent) and flooding (41.5 per cent), were the main disaster events of the losses 1990-2014 (Li *et al.*, 2016).

Several trends in weather extremes are sufficiently clear to inform the potential increases in extreme events due to climate change come on top of alarming rises in vulnerability. The additional risks due to climate change should not be analyzed or treated in isolation, but instead integrated into broader efforts to reduce the risk of natural disasters. Climate change is one of the most complexes, multifaceted and serious threats the world faces. The response to this threat is fundamentally linked to pressing concerns of sustainable development and global fairness; of economy, poverty reduction and society; and of the world we want to hand down to our children."

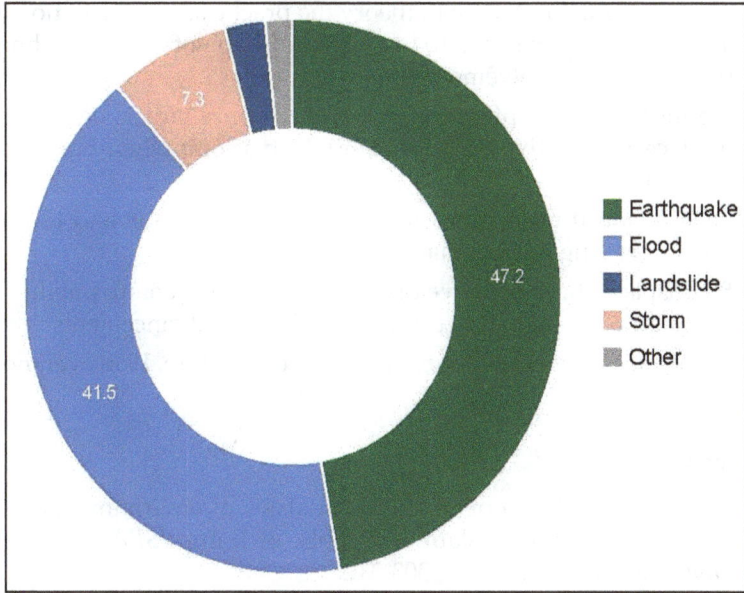

Figure 3.6: Frequency of the International Disaster during 1990–2014.

Carbon dioxide (CO_2) is the primary greenhouse gas emitted through human activities. In 2013, CO_2 accounted for about 82 per cent of all U.S. greenhouse gas emissions from human activities. Carbon dioxide is naturally present in the atmosphere as part of the Earth's carbon cycle (the natural circulation of carbon among the atmosphere, oceans, soil, plants, and animals). Human activities are altering the carbon cycle—both by adding more CO_2 to the atmosphere and by influencing the ability of natural sinks, like forests, to remove CO_2 from the atmosphere. While CO_2 emissions come from a variety of natural sources, human-related emissions are responsible for the increase that has occurred in the atmosphere since the industrial revolution.

Acknowledgements

The authors are honored to be one of the recipients of the grant to take part in the international workshop on mitigation of disaster due to sever climate events, from policy to practice, Colombo, Sri Lanka, 10-13 March 2016. Thanks to NAM S&T Centre and generous support from the Sri Lankan hosts. As the main objective of the workshop is to bring the stakeholders at national and international levels together, with participation of learned experts, to share experience and work together in a scientific manner for working out strategic interventions to mitigate the hazards due to severe climate events. We can make climate information "usable" by the public, by policy makers and by businesses. It is hoped that the outcomes could clarity the scientific outlook and practical solutions on climate change phenomenon and natural disaster more than ever. At the end some important suggestions are proposed for consideration:

- ☆ Though a great initiative in theory the practicality of execution is absent as the approach taken is to top down and do not involve technocrats or private sector involvement.

- ☆ No mechanism in place to track and measure targets and no penalties in place as incentives to do so. For which a comprehensive legislation is required.

- ☆ Lack financial mechanism to ensure access to funds and technological transfer among most climate effected countries.

- ☆ Big disparity between developed and developing country ability to adapt and regularize knowledge and technological advancements.

- ☆ These ideas have been touted around years before. However, nothing or little has changed in reality.

References

1. Avami, A., Farahmandpour, B., 2008. Analysis of environmental emissions and greenhouse gases in Islamic Republic of Iran. *WSEAS Transactions on Environment and Development* 4, 303-312.

2. Li, C.-j., Chai, Y.-q., Yang, L.-s., Li, H.-r., 2016. Spatio-temporal distribution of flood disasters and analysis of influencing factors in Africa. *Natural Hazards*, 1-11.

3. Madani, K., 2014. Water management in Iran: what is causing the looming crisis? *Journal of Environmental Studies and Sciences* 4, 315-328.

4. O'Neill, B.C., Oppenheimer, M., 2002. Dangerous climate impacts and the Kyoto Protocol. *Science* 296, 1971-1972.

5. Roshan, G., Yousefi, R., Fitchett, J.M., 2016. Long-term trends in tourism climate index scores for 40 stations across Iran: the role of climate change and influence on tourism sustainability. *International Journal of Biometeorology* 60, 33-52.

6. Roshan, G.R., Lagh, F.K., Azizi, G., Mohammadi, H., 2011. Simulation of temperature changes in Iran under the atmosphere carbon dioxide duplication condition. *Iranian Journal of Environmental Health Science and Engineering* 8, 139.

7. Shahbazi, H., Reyhanian, M., Hosseini, V., Afshin, H., 2016. The Relative Contributions of Mobile Sources to Air Pollutant Emissions in Tehran, Iran: an Emission Inventory Approach. *Emission Control Science and Technology* 2, 44-56.

8. Smith, P., Davis, S.J., Creutzig, F., Fuss, S., Minx, J., Gabrielle, B., Kato, E., Jackson, R.B., Cowie, A., Kriegler, E., 2016. Biophysical and economic limits to negative CO_2 emissions. *Nature Climate Change* 6, 42-50.

9.   Tofigh, A.A., Abedian, M., 2016. Analysis of energy status in Iran for designing sustainable energy roadmap. *Renewable and Sustainable Energy Reviews* 57, 1296-1306.

10. Wilhite, D.A., Sivakumar, M.V.K., Pulwarty, R., 2014. Managing drought risk in a changing climate: The role of national drought policy. *Weather and Climate Extremes* 3, 4-13.

Chapter 4

Challenges in the Implementation of Climate Change Policy of Nepal

Jiba Raj Pokharel

Vice Chancellor,
Nepal Academy of Science and Technology,
Kathmandu, Nepal
E-mail: jibaraj@gmail.com

ABSTRACT

Nepal with 0.4 per cent of the world population contributes to a mere 0.025 per cent of the global greenhouse gas emission. Still, with the increase in temperature of around 0.06 degree centigrade per year, it is very vulnerable to climate change in view of vulnerability projection partly because of the location of more than 2000 glaciers in its north due to constant ice melt. In order to attract the attention of the whole world, it had held the cabinet meeting in Kala Pathhar Mountain near the base camp of the Mountain Everest, the tallest in the world with the height of 8848 meter before the commencement of the Copenhagen conference in 2012. Correspondingly, it has prepared a Climate Change Policy in 2011 with the objectives of firstly, reducing GHG (GREEN HOUSE GAS) emissions by promoting the use of clean energy; secondly, enhancing the climate adaptation and resilience capacity of local communities for optimum utilization of natural resources and their efficient management; and thirdly, adopting a low-carbon development pathway by pursuing climate- resilient socio-economic development. But it has yet to materialize several of the provisions mentioned in the policy. This paper seeks to highlight on the challenges that are likely to be faced in the execution of all the aspects of the policy.

Introduction

That the whole of the earth has been subjected to steady warming has now received widespread acceptance round the globe. The Paris conference held earlier this year has deliberated on this problem and it has also arrived at a global commitment to confine the warming to 1.5 degree centigrade to the extent possible and not exceed 2 degree centigrade in any circumstances. For this a ten point plan of action also has been formulated and the nation states have committed to follow it seriously.

Nepal cannot be an exception to this universal reality. Though it has 0.4 per cent of the world population of it merely contributes to 0.025 per cent in carbon emission. The overall temperature has increased by 1.8 degree centigrade between 1975 and 2006, while the high altitude area of the country is experiencing an increase of 1.2 degree centigrade (GoN, 2012). Due to this, Nepal is very vulnerable to climate change mainly because of the location of Himalayas, the ice capped mountains including the tallest of the world the Mount Everest with a height of 8848 meters. It has already experienced an average change in temperature of 0.06 degree centigrade. According to German Watch 2006, it is the sixth most vulnerable country to climate change induced disasters.

Besides there are more than 2000 glaciers in this elevated area of which many are vulnerable to melting due to increase in annual temperature due to global warming.

Climate Change can also trigger the collapse of the ruling dynasties. Zhang *et al.,* describe how the Ming Dynasty in China(1368-1644) was defeated by the following Quing Dynasty (1644-1911) due to the climate change occurring in China during that period. In Nepal also this period was marked by drought and famine.

Nepal's Efforts to Combat Climate Change in the Past

The International Panel on Climate Change (IPCC) had long been attracting the attention of the world towards the global warming emerging steadily and strongly since its inception in 1988. As a result, General Assembly of the United Nations adopted a resolution towards framing a legal instrument. The Inter-Governmental Negotiation Committee decided to agree on a general framework to address climate change following which the United Nations Framework Convention on Climate Change (UNFCC) was adopted. It was followed by the organization of the UN Conference in Rio De Janerio, Brazil where this treaty was open for ratification. Nepal became a party to it by being a signatory on 12 June 1992.

After being a party to the General Convention, Nepal prepared an Initial National Communication in 1994 which was then shared with the Convention Secretariat. Then the Ministry of Population and Environment, now Ministry of Environment, was designated as the focal point to implement the commitments made in UNFCC. It was also given the responsibility to promote the Projects of Clean Development Mechanism. In this connection, some public awareness programs were also organized.

Nepal later carried out some important works related to climate change, which are:

☆ Preparation of the Action Plan under the National Capacity Needs Assessment Project for capacity building

☆ Initiation of action towards the CDM project as per the Kyoto Protocol

☆ Beginning of the Preparation of National Adaptation Plan of Action

☆ Preparation of the Second National Communication

☆ Implementation of a Project on the strengthening the capacity for Climate Change management

☆ Holding of the Cabinet Meeting in Kala Pathhar near the Base Camp of Mount Everest

☆ Organization of a Regional Seminar on Kathmandu to Copenhagen

☆ Preparation of Status Paper for COP 15

☆ Formation of a 15 member Climate Change Council under the Chairmanship of the Prime Minister

☆ Inscription of the Environment and Climate Change agenda in the Constitution of Nepal

Climate Change Policy of Nepal

Nepal later prepared a Climate Change Policy on the virtue of being a party to decisions of the COP held in Marrakesh in the year 2010. It is primarily because interest is emerging around the world to design and implement mandatory domestic climate based climate change policies (Aldy *et al.*, 2008). Policies on climate change can help to achieve such objectives such as enhanced energy security and environmental protection targets. Moreover, OECD analysis reveals that CHG emissions reductions are achievable at low cost if the right Policies are in place (OECD, 2007). Policy is generally understood as a guiding gazette document which is prepared with objectives and prescriptions, instructions and targets to meet the objectives (Helvetas, 2011). Policies should generally be a product of the domestic need. But Nepal's climate change policy has emerged due to the international commitment arising out of the vulnerability of Nepal to Climate Change. For example Maplecroft has put Nepal in the range of the fourth most vulnerable country to climate change after Bangladesh, India and Madagascar.

This Policy is outlined in fourteen different sections. The first section describes the background while the second section focuses on the past efforts of Nepal towards addressing the climate change. The third section deals with the present situation while the fourth section highlights the problems and challenges. Vision and Mission have been put forward in the fifth section with the seventh section concentrates on the goals and the targets. The more important objectives have been bracketed under the eighth section while the following ninth section declares the strategy and the working policy. It then leads to the tenth section describing the institutional structure. Legal aspects and monitoring are included in the eleventh, twelfth and thirteenth section. It finally comes to an end with the Risk mitigation/reduction aspects in the fourteenth section.

The objectives of the Policy are as follows:

☆ To establish a Climate Change Center as an effective technical institution to address issues of climate change and also strengthen existing institutions

☆ To implement climate adaptation related programs and maximize the benefits by enhancing positive impacts and mitigating the adverse impacts

☆ To reduce GHG emissions by promoting the use of clean energy, such as hydro- electricity, renewable and alternative energies, and by increasing energy efficiency and encouraging the use of green technology

☆ To enhance the climate adaptation and resilience capacity of local communities for optimum utilization of natural resources and their efficient management

☆ To adopt a low carbon development path by pursuing resilient socio economic development path

☆ To develop capacity for identifying and quantifying present and future impacts of climate change, adapting to climate risks and adverse impacts of climate change and

☆ To improve the living standard of people by maximum utilization of the opportunities created from the climate change related conventions, protocols and agreements

Establishment of a Climate Change Center

There are several Climate Change Centers established round the globe. One of them is the National Climate Change and Wild Life Science Center in the United States with its office in Virginia. It has Climate Change Centers located in eight places. It mainly delivers basic climate change science impacts; prioritizes fundamental science, data and decision support activities; partners and helps facilitate the coordination of fundamental climate science capabilities across their region of responsibility; synthesizes, integrates and communicates existing climate change impact data; partners with resource managers at pertinent LCCs to assist development of science based adaptation strategies.

In Nepal, the Nepal Climate Change Knowledge Management Center (NCCKMC) has been established by a joint effort of Nepal Academy of Science and Technology and the Ministry of Environment. As is obvious from it name, it has a mission to operate as a dedicated institutional arrangement for managing climate change knowledge in Nepal. Besides, it aims to offer service as an arena for coordinating and facilitating the regular generation together with the management and exchange as well as dissemination of climate related knowledge and services catering to a multi stakeholder community in Nepal. Its objectives are:

☆ To build climate change learning in Nepal particularly

☆ To enhance public access to climate change and related information in order to build their capacities to address the challenges posed by climate change

☆ To capture learning objectives and best approaches

The deliveries are in the form of capacity building, climate change learning for young researchers, learning events/seminars for policy makers, popularization and translation, mobile libraries and communication strategies.

Implementation of Climate Change Adaptation Related programs

Nepal has prepared a National Adaptation Plan of Action (NAPA) after a rigorous consultation with multiple stakeholders referring to the annotated guidelines developed by the Least Developed Countries Expert Group in September 2010. The objectives were:

☆ Assessing and prioritizing climate change vulnerabilities and identifying adaptation measures

☆ Developing proposals for priority activities

☆ Preparing, reviewing and finalizing NAPA focus areas

☆ Developing and maintaining a knowledge management and learning platform

☆ Developing a multi stakeholder framework for action on Climate Change

Nepal has also prepared Local Adaptation Plan of Action (LAPA) and Community Adaptation Plan of Action (CAPA). At present, Nepal is executing LAPAs in 90 Village Development Committees and 7 Municipalities – the lowest administrative units in the country. Similarly, about 375 local adaptation plans and nearly 2200 community adaptation plans of action (CAPAs) for community forests have been developed.

Reduction of GHG Emissions through the Use of Clean Energy

In this case, the direct and indirect GHG emissions have to be taken in account (Scmidt, 2008) which is then planned to be tackled through several measures. Traditional sources like bio mass form the brunt of Nepali energy usage. Renewable energy accounts for only 1 per cent of the total energy consumption which the Government plans to increase it to 20 per cent by 2020. This is despite the high potential of hydropower, solar power and wind energy present in the country. It remains unharnessed primarily because of geographically inaccessible, political and economic reasons. Consequently, the country is facing a load shedding of up to 15 hours per day. But Nepal has been executing National Rural and Renewable Energy Program since almost two decades now through the promotion of micro hydro, solar, biogas and improved cook stoves. This plan has now been extended to the urban areas also. The renewable energy technologies have received a boost due to the fuel crisis following which these technologies are being increasingly used in Nepal.

The forest cover in Nepal is believed to be about 30 per cent. The country has however the strategy to maintain at least 40 per cent of the total area of the country under the forests. This it aims to achieve through the afforestation in public and private lands, promoting environment friendly infrastructure development followed by the conservation of biodiversity. The country has a plan to promote the management of ecosystems in the different eco regions of the country, the Himalayas, the Hills and the Plains through the sustained management of the forests, the enhancement of the capacity of the local communities in adaptation and resilience. The achievement of carbon storage through sustainable forest management is another national effort towards the reduction of carbon emissions. In this context, it has been announced The Forest Decade for 2014-2023 with the slogan, one house one tree, one village one forest and one town with several parks. It is coupled with the management of the forest and watersheds lying on the chain of Siwalik hill on the basis of upstream- downstream linkages. This is expected not only to sequestrate carbon but also function as carbon sink.

The other attempt to reduce GHG is through the execution of an Environment Friendly Vehicle and Transport Policy (2014) by increasing the share of the electric vehicle to 20 per cent by 2020 through the provision of subsidy scheme for the promotion of electric and non- motorized vehicles. It is because the Transport Planning also contributes to the reduction of GHG emissions (Minoia *et al.*, 2008). Avoidance of the unnecessary travel, reduction of trip distance and promotion of the shift towards more sustainable transport modes such as non- motorized transport form the hall mark of the transport plan. This policy has focussed on improvement of the transport practices and technologies by diversifying towards electricity for fuel quality along with the regulation of vehicle emissions in an effort to ensure compliance with air quality. Emphasis on public transport systems and the increased use of bicycles along with the beginning of fuel tax in Kathmandu for air quality improvement is expected to reduce the pollution in urban areas.

To Enhance the Climate Adaptation and Resilience Capacity of the Communities

Nepal has stressed on the need to mitigate adverse impacts of climate change through the adaptation measures such as the mitigation of friendly forest management systems. These consist of making community- based forests and watershed management climate adaptation-friendly, enhancing carbon sequestration through sustainable management of forests, and supporting programmes that reduce carbon emissions from forest areas. More than 25,000 community-based forest management groups in the country are geared towards managing about 34 per cent of the country's total forest area. It has resulted in sequestering carbon dioxide by sustainable management of forest resources simultaneously also playing effective roles in designing and implementing community adaptation plans of action (CAPAs) based on forests and non-forests benefits. Based on climate change mitigation and resilience, the Forestry Sector Strategy (2016-2025) aims to enhance Nepal's forest carbon stock by at least 5 per cent by 2025 as compared to 2015 level, as well as to decrease mean annual deforestation rate by 0.05 per cent from about 0.44 per cent and 0.18 per cent in the Terai and Siwalik hills respectively(GoN, 2016). The aim is also to put in place forest carbon trade and payment mechanism and mainstream community/ecosystem-based adaptation by 2025. Forest areas are sought to be managed in a variety of ways including community forests, leasehold forests, collaborative forests and protected areas pursuing a landscape approach to resource conservation and management.

The Challenges to Implement the Policy

The challenges have been mentioned in the policy itself and these consist of the following:

☆ The lack of knowledge, scientific data, and information related to the science of climate change and its impact on different geographical and socio economic development sectors and use of climate modeling to assess likely impacts

Nepal consists of high altitude mountainous areas. There is lack of data series required for flood hazard analysis, risk assessment and mitigation strategies. For example, the fluvial variability in the Alps was found out by making use of different methods from several disciplines such as sedimentology, geomorphology, geography, palynology and geochronology. A 3600 year flood history was derived delta proxies which was unique in Europe Alps (Schulte *et al.*, 2009). This kind of information is not available in Nepal. Nepal as in the case of other countries as well suffers from lack of qualified personnel and documentation (Schulte *et al.*, 2009). This is certainly one of the challenges.

☆ to assess the effects and likely impacts of climate change to identify the vulnerable sectors and enhance their adaptive capacity, and to develop a mechanism for reducing GHG emissions

The vulnerability analysis of all the 75 districts in Nepal has been carried out according to which the districts are classified as very high, high, moderate, low and very low. Kathmandu, Bhaktapur and Lalitpur rank as very highly vulnerable and have been recommended for mitigation measures. But the report concedes that it is suffering from the lack of data saying *there is always room for improvement of the sensitivity, risk/exposure and adaptation capability indices, and consequently the map prepared on the basis of above indices* (GoN, 2010).

The reduction of GHG emissions have been attempted by the geological storage of Carbon Di oxide in Alberta Canada (Gunter *et al.*, 2009) but such a sophisticated method is difficult to be followed in the case of Nepal. It has to rely on simple methods.

☆ To create an enabling environment for technical and financial opportunities at the national and international level in the process of addressing climate change impacts

The Policy has talked of the establishment of Climate Change Fund for mobilizing the financial resources from public and private, internal and external sources to address the issues of climate change. The country has also made substantial investment as reflected by the expenditure of 2, 1.78 and 3.1 per cent of GDP in the years 2011, 2012 and 2013. It is 7.2, 6.74 and 10.2 per cent of the national budget of those years. (GoN, 2013). The Policy has also committed to allocate at least 80 per cent of the available funds for field level climate change activities.

It has aimed to generate financial resources by promoting carbon trade and Clean Development Mechanism. It also seeks to obtain resources through the implementation of the polluter pays principle.

Several donors such as Asian Development Bank (ADB), Denmark, European Union (EU), Finland, Germany, Japan, Norway, Switzerland, United Kingdom (UK), United Nations (UN) and the World Bank have contributed with climate change activities in Nepal. They invested $652.4 million in the year 2014with the World Bank topping the list with a contribution of $ 379.4 million (GoN, 2011-12).

☆ To make the country's socio economic development climate friendly, and to integrate climate change aspects into policies, laws, plans and development programs and implement them

There is a need of developing local plans because these contribute to the reduction of global emissions (Tang *et al.*, 2010). There is also a need to mainstream climate change management in long term plans and policies and in particular to integrate climate risks and risks reduction effort into periodic development plan (GoN, 2012).

☆ To establish the current and likely adverse impacts of climate change between upstream and downstream areas so as to promote regional cooperation

☆ To effectively enhance the capacity of public institutions, planners and technicians, private sectors, NGOs, and civil society involved in the development work

☆ To give attention to develop a capable organizational structure with necessary financial and human resources for addressing climate change issues

Potential financial resources are funds under the UNFCCC, the global environmental facility, non-compliance fund, disaster relief and risk reduction, public expenditures including public private partnerships, insurance and disaster pooling, development assistance and foreign direct investment (Bouwer *et al.*, 2006).

☆ To take full advantage of the international climate change regime in order to achieve the UN Millennium Development Goals and avoid or minimize the impacts of climate change on mountainous environments, people and their livelihood and ecosystems

Conclusions

It has been found that though the political regime is not the determinant for a successful climate change policy implementation, the public knowledge of climate change is a strong determinant (Steves, 2013). Public knowledge is shaped among other things, chiefly by the threat posed by climate change, the national level of education and the national media. For Nepali people, the climate change is not the main worry. 30 per cent of Nepalese worry for not having enough food to eat, 29 per cent for not being able to send the children to School, 13 per cent for not having enough clean water to drink, another 13 per cent for not being healthy, 8 per cent for not having electricity, 5 per cent for not having a suitable house, 3 per cent for not having money to buy suitable items and 1 per cent for not being able to buy a latest model of mobile phone(Colon Anna *et al.*, 2013). Directly, the Climate Change thus does not appear as the primary threat even though people have perceived extreme change in weather.

1. The literacy in Nepal is about 65.9 per cent (2011 National Census), it being more among the males (75.1 per cent) and less among the females (57.4 per cent). It may be one stumbling block for the successful implementation of the Climate Change Policy.

2. 70 per cent of the people having high exposure to social communication media are concerned about the climate change while only 51 per cent of them having no exposure worry about it (Colon Anna *et al.*, 2013).

3. All these indicators present not a very smooth trajectory for the implementation of the Climate Change Policy for Nepal.

References

1. Aldy Joseph E *et al.*, Issues in designing US Climate Change Policy, Resources for the Future.

2. Bouwer *et al.*, Financing Climate Change Adaptation, Disasters.

3. Colon Anna *et al.*, 2013, Nepal, Climate Asia.

4. Government of Nepal (GoN), 2010, Climate Change Vulnerability for Nepal, Ministry of Environment.

5. Government of Nepal, 2011, Climate Change Policy, Ministry of Environment.

6. Government of Nepal, 2012, Resilient Planning, National Planning Commission.

7. Government of Nepal, 2013, Climate Change Budget Code, National Planning Commission.

8. Government of Nepal, 2016, Intended Nationally Determined Contributions, Ministry of Population and Environment.

9. Gunter W D *et al.*, 2009, Do high resolution fan delta records provide useful tool for hazard assessment in mountain regions ? *International Journal of Climate Changes and Management*, Vol. 1 No. 2, 2009, pp. 160-178, Emerald Group Publishing Limited, 1756-8692.

10. Helvetas, 2011, Nepal's Climate Change Policies and Plans, Swiss Interco-operation.

11. OECD, 2007, Policy Briefs, Public Affairs Divisioin.

12. Maris Klavins, Ieva Bruneniece and Valdis Bisters, 2008, Development of National Climate and Adaptation Policy in Latvia, *International Journal of Climate Changes and Management*, Vol. No. 1, 2009, pp. 75-91, Emerald Group Publishing Limited, 1756-8692.

13. National Planning Commission(NPC), 2011, The Future of Climate Finance in Nepal, Government of Nepal.

14. Paulo Minoia and Gabriele Zanetto, 2008, An assessment of principle of subsidiary in urban planning to face climate change, *International Journal of Climate Changes and Management*, Vol. No. 1, 2009, pp. 63-74, Emerald Group Publishing Limited, 1756-8692.

15. Schmidt Mario, 2008, Carbon accounting and carbon footprint- more than just diced results?, *International Journal of Climate Changes and Management*, Vol. No. 1, 2009, pp. 19-30, Emerald Group Publishing Limited, 1756-8692.

16. Schulte L *et al.*, 2009, Do high resolution fan delta records provide useful tool for hazard assessment in mountain regions, *International Journal of Climate Changes and Management*, Vol. 1 No. 2, 2009, pp. 197-210, Emerald Group Publishing Limited, 1756-8692.

17. Steves F, 2013, Political Economy of Climate Change Policies in the Transition Region, European Bank for Reconstruction and Development.

18. Tang *et al.*, 2010, Measuring local climate change response capacity and bridging gaps between local action plans and land use plans, *International Journal of Climate Changes and Management*, Vol. 3 No. 1, 2009, pp. 74-100, Emerald Group Publishing Limited, 1756-8692.

19. Walter Leal Filho, 2009, Communicating Climate Change: Challenges ahead and Actions needed, *International Journal of Climate Changes and Management*, Vol. 1 No. 2, 2009, pp. 6-18, Emerald Group Publishing Limited, 1756-8692.

20. Zhang *et al.*, 2014, How Climate Change Impacted the fall of Ming Dynasty, Climate Change, Springer.

Chapter 5

Mitigating the Effects of Adverse Climatic Conditions in Zambia

Shadreck Mpanga[1]*, Mwansa Kaoma[1]**,
Kennedy Zimba[2] and Ackim Zulu[1]***

[1]University of Zambia,
School of Engineering, Electrical Engineering,
P.O. Box 32379, Lusaka, Zambia
E-mail: *shadreck.mpanga@unza.zm; **mwansa.kaoma@unza.zm;
***ackim.zulu@unza.zm
[2]University of Zambia,
School of Agricultural Sciences, Agricultural Sciences,
P.O. Box 32379, Lusaka, Zambia
E-mail: kennedy.zimba@unza.zm

ABSTRACT

Zambia possesses about 40 per cent of the water resources in Southern Africa. However in recent years it has seen a reduced rainfall activity. This situation has led to the low water levels in two major water reservoirs used for hydroelectric power generation at the Kafue Gorge Power Station and the Kariba North Bank Power Station. As nearly 95 per cent of the generated electricity comes from the hydro resources, reduced rainfall activity has led to power rationing recently to a level which has never been experienced before. The 2015-2016 raining season, which runs from late October to early April, is forecasted to be even worse due to the El Niño effect, with resulted reduced rainfall over much of Southern Africa. In addition, the 2015 hot season turned out to be the hottest season Zambia has ever experienced, and led to high evaporation losses from open surface water bodies. This paper discusses the approach, in terms of policy and practice, the Zambian authorities have taken to mitigate the effects of the adverse climatic conditions in the energy and agricultural sectors of the country. The success of large scale farming is dependent on sustainable supply of electricity for irrigation. The sectors of agriculture and energy are currently under threat in Zambia due to the drought experienced in the 2015-2016 rain season. To survive, the country is looking at alternative approaches of sustaining these important sectors of economy. Some of the measures taken are investing more resources into water infrastructure and water harvesting because Zambia has numerous water bodies.

Keywords: Energy, Agriculture, Climate change, Risk mitigation, Policy, Practice.

Introduction

About 40 per cent of the water resources in Southern Africa are found in Zambia. This explains why 95 per cent of generated electricity in Zambia is from hydro related resources. Zambia has seen a reduced rainfall activity in recent years and this has led to the reduced water levels in the two major reservoirs used for hydroelectric generation at Kafue Gorge power station and Kariba North Bank power station. The reduced rainfall activity (rain season in Zambia runs from late October to early April) has led to load shedding, the level of which has never been experienced before in the history of Zambia. For instance, as of December 2015, out of the installed capacity of about 2300 MW, Zambia could produce only 1100 MW against a national demand of 1600 MW. The 2015-2016 rain reason has been worse due to the El Niño effect and this has led to suppressed rainfall over much of Southern Africa (WFP, 2015). In addition, the 2015 hot season proved to be the hottest season Zambia has ever experienced, which led to high evaporation losses from open surface water bodies.

Due to this severe climatic situation, two sectors of economy, energy and agriculture, are particularly at risk. The response of the Zambian authorities is to put in place measures to mitigate the effects of these adverse climatic conditions in the energy and agriculture sectors of the country. The success of large scale farming is dependent on sustainable supply of electricity for irrigation. It is imperative that the authorities need to revisit the policy on energy and agriculture and redirect to cope with this kind of crisis. The policies must guide the country to look to other means of sustaining these important sectors of economy. With this situation in mind, Zambia has developed the following policies.

Energy Policy

Excluding petroleum, which is wholly imported, Zambia is endowed with plenty of indigenous energy resources such as woodlands for wood fuel, hydropower, coal and renewable energy. The country's energy policy encourages diversification of the country's energy mix through use of Renewable Energy, creating conditions that ensure availability of adequate supply of energy from various sources which are dependable at lowest economic, financial, social and environmental costs consistent with national development goals (ZEP, 2008).

Agriculture Policy

The agriculture policy for Zambia aims at reducing poverty through broad-based income growth in the agricultural sector, attaining 90 per cent household food security and cutting hunger by 50 per cent by 2015, growing the agriculture sector from 1 per cent to 7-10 per cent per year, increasing agricultural contribution to GDP from 18-20 per cent to 25 per cent and increasing agriculture contribution to foreign currency earning from 3-5 per cent to 10-20 per cent (Kuteya, 2012).

Research Policy

Research will help to understand and adapt to climate change and the Government through the Ministry of Higher Education has sanctioned the following actions:

(a) In the Energy Category, topics on development and adaption of technologies for renewable energy and improving the performance and efficiency of renewable energy sources have been chosen.

(b) In the Agriculture category, topics on improving the productivity of communally grazed range lands and development of maize varieties that adapt to changing climate are being pursued. Maize is the staple food in Zambia.

The measures in research will help the Government in mitigating the disasters arising from severe climate events.

One center that oversees the research in the country is the Zambia Institute for Policy Analysis and Research (ZIPAR), a semi-autonomous think-tank involved in policy formulation, implementation and monitoring. It was established by the Government Republic of Zambia (GRZ) with the support of the African Capacity Building Foundation (ACBF) in 2009. The Institute supports the Government, the private sector, civil society and other stakeholders in Zambia on evidence-based policy through conducting primary and secondary research (ZIPAR, 2015).

Overview of the Energy Sector in Zambia

Energy Demand

Energy is the primary driver of any economic activity in a country. As is seen in Figure 5.1, the footprint of electrical energy is small (around 14 per cent) while the use of wood fuel exerts enormous influence on the energy used in Zambia. Of the electrical energy in use, mining activities take the largest chunk (more that 50 per cent) as shown in Figure 5.2.

There is considerable scope to develop the electrical energy sector. As shown in Figures 5.3 and 5.4 there are more prospects for electrical energy generation. This situation gives hope of a solution to the current loading shedding taking place in Zambia.

Energy Reforms

When Zambia got independence from Britain in 1964, the electrical energy infrastructure development started with the Victoria falls, Kariba North Bank and Kafue Gorge power stations whose installed capacities are shown in Table 5.1. These are the three main generation stations in Zambia.

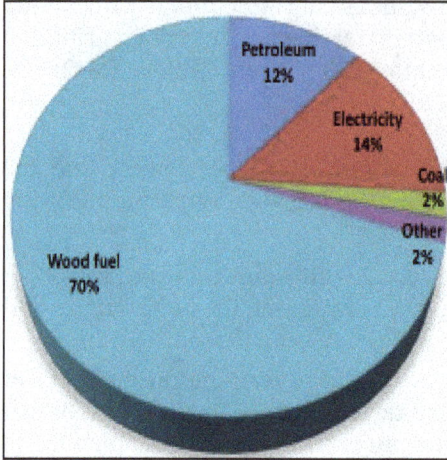

Figure 5.1: Energy Demand by Source (Muzeya, 2015).

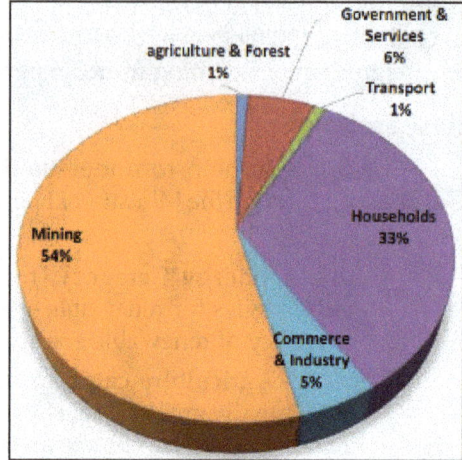

Figure 5.2: Electricity Consumption by Group (Muzeya, 2015).

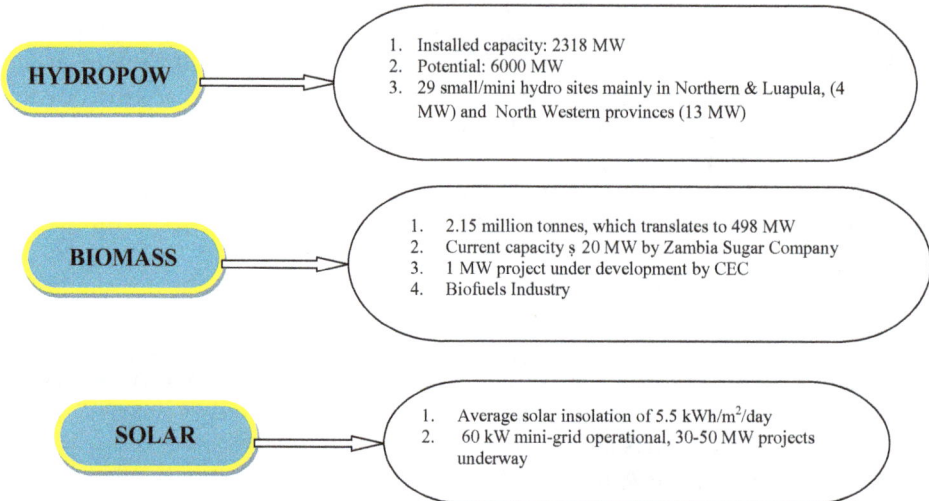

HYDROPOW

1. Installed capacity: 2318 MW
2. Potential: 6000 MW
3. 29 small/mini hydro sites mainly in Northern & Luapula, (4 MW) and North Western provinces (13 MW)

BIOMASS

1. 2.15 million tonnes, which translates to 498 MW
2. Current capacity ş 20 MW by Zambia Sugar Company
3. 1 MW project under development by CEC
4. Biofuels Industry

SOLAR

1. Average solar insolation of 5.5 kWh/m^2/day
2. 60 kW mini-grid operational, 30-50 MW projects underway

Figure 5.3: Energy Sources Already Explored (Muzeya, 2015; ESP, 2014).

Table 5.1: Zambia's Installed Hydro Generation Capacity (Muzeya, 2015; ESP, 2014)

Item	Power Station	Original Capacity [MW]	Upgraded Capacity [MW]
1.	Kafue Gorge	900	990
2.	Kariba North Bank	600	1080
3.	Victoria Falls	108	108

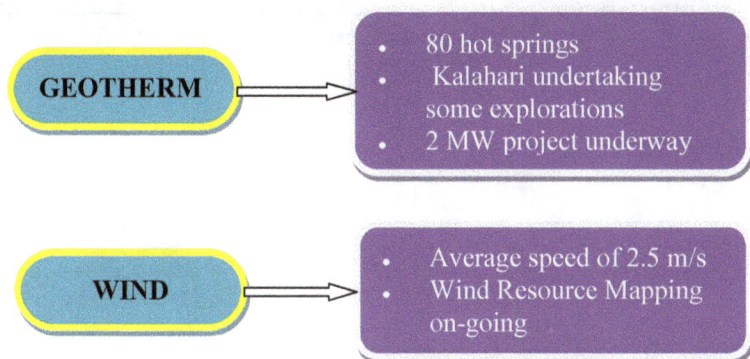

**Figure 5.4: Energy Sources Currently under Exploration
(Muzeya, 2015; ESP, 2014).**

In 1970, there was nationalization of the power infrastructure and Zesco was the only company to generate, transmit and distribute electricity. In 1994, the Energy policy was developed in which there was a provision of the Rural Electrification Fund (Muzeya, 2015). In 1995, there was a liberisation of the energy sector when the Electricity Act (Muzeya, 2015) and Energy Regulation Act (Muzeya, 2015) were instituted. The enactment of Rural Electrification Act (Muzeya, 2015) in 2003 facilitated the establishment of Rural Electrification Authority (Muzeya, 2015) in the same year. The Vision 2030 (Muzeya, 2015; ESP, 2014) was formulated in 2006 and the Energy Policy was revised in 2008. The Vision 2030 aims at increasing access to electricity by the year 2030. The national target is 66 per cent with the urban target being 91 per cent and rural 51 per cent. The Power System Development Master Plan (Muzeya, 2015) developed in 2008 aims at increasing the total generated capacity of electricity to 4337 MW by the year 2030. The Energy Policy has recognized coal as a source to contribute to the national Energy mix and as such 300 MW of a coal power plant is under development, with 150 MW likely to be commissioned by mid 2016. All these measures will help reduce dependence on wood fuel and ensure sustainable provision of affordable, reliable modern energy services to rural and urban households as a means of raising productivity and standards of living (Muzeya, 2015; ESP, 2014).

Institutional Framework

Figure 5.5 shows how energy policies are formulated and how they are implemented in Figure 5.6.

Current Energ Deficit

Table 5.2 shows the generation of electricity from the three main generation stations as of December 2015.

In February, 2016, the Minister of Energy and Water Development informed parliament that the country still had an energy deficit of 1000 MW. Table 5.2 clearly shows how the capacities of the power stations are unused. From this scenario lies the basis for the current load shedding in the country. The discrepancy between

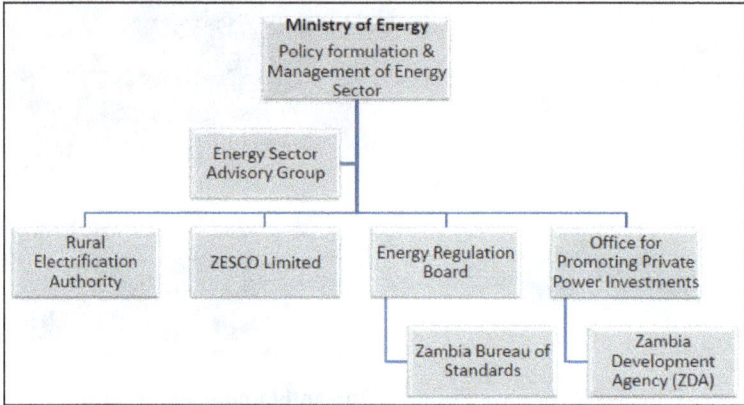

Figure 5.5: Institutional Framework (ESP, 2014).

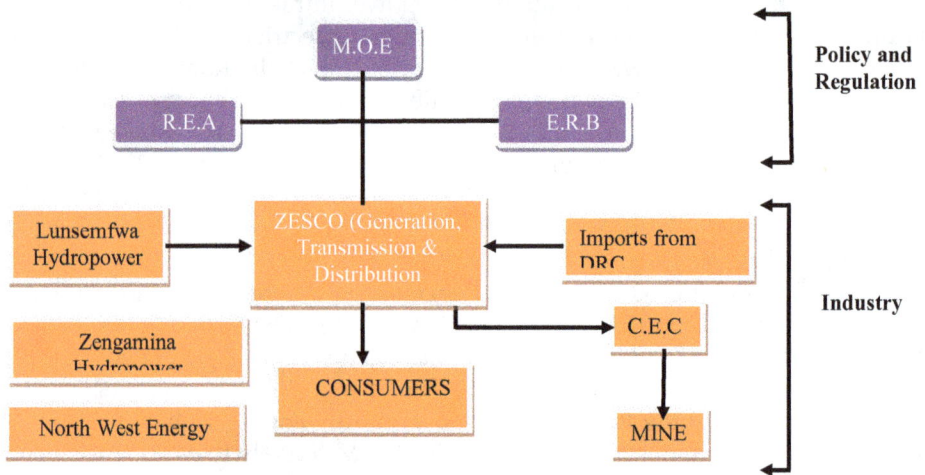

Figure 5.6: Policy Implementation (ESP, 2014).

the installed capacities and the generated amounts, particularly for Kariba North Bank, is due to low water levels in the reservoirs. Table 5.3 shows a downward trend in the amount of power output from the stations, with Victoria Falls being the only station showing an upward trend from December 2015 to February 2016.

Table 5.2: Generated Electricity from Three Main Stations

Item	Power Station	Amount Generated [MW]	Installed Capacity [MW]
1.	Kariba North Bank	310	1080
2.	Kafue Gorge	650	990
3.	Victoria Falls	55	108
	TOTAL	**1015**	**2178**

Table 5.3: Recommended Generation in the Month of February 2016

Item	Power Station	Generation Amount [MW]	Installed Capacity [MW]
1.	Kariba North Bank	275	1080
2.	Kafue Gorge	540	990
3.	Victoria Falls	91	108
4.	Itezhi-tezhi (New)	30	60 (×2 later)
5.	Lunsemfwa	14	56
6.	Small Hydros	13	25
	TOTAL	963	2319

Low water levels in the reservoir at Kariba led to water rationing by the water regulatory body such as Zambezi River Authority. The trend from December to February as depicted in Tables 5.2 and 5.3 shows a critical rationing routine. If power generation stations are allowed to run at full capacity, the water reservoirs would dry up in a matter of months. The situation of diminished generated power has affected industries, commerce, agriculture and households. The sad effects of this have been the job losses in the mines and other sectors of the economy as the companies cannot operate at their capacities. The reduced rainfall activity has also led to a section of society getting concerned about the agricultural produce later this year. Some other sections have already sounded alarm that there is a looming food crisis in the country.

Mitigation Measures

There are immediate and long-term actions that can be applied to address the energy and food shortages that are resulting from the adverse effects of climate change. Both government and the private sector, together with nongovernmental organizations could be involved. A discussion of these measures as applied in Zambia follows below.

Energy Solutions

To mitigate the power shortages, the Government intends to bring in about 300 MW of electricity from a coal fired power plant that will be commissioned in mid 2016. Government has also signed an MOU with South Africa to start importing about 300 MW. In addition, the Government has already shortlisted some companies, both national and international, to install large Photovoltaic (PV) power generating systems to be connected to the grid. The target capacity for the PV plants at the moment is about 300 MW in the next few years. In the meantime, the Government intends to produce about 50 MW of solar power and inject it into the main grid in the next few months. The other big projects being undertaken by the government to solve the energy crisis are:

    (a) 120 MW Itezhi Tezhi hydro power with 30 MW already being delivered as of early February, 2016;

(b) Batoka Gorge with a capacity of about 1, 600 MW, with pre-feasibility studies already done;

(c) Kafue Lower Hydropower Plants with expected generation capacity of 750 MW.

In addition to the above measures, the government has proposed the creation of at least two water reservoirs up the Zambezi River that can later release water into the Kariba dam in dry months of the year. A more complete picture of the total installed power generation capacity will look as shown in Table 5.4.

Table 5.4: Expected installed generation by 2030

Item	Power Station	Type	Installed Capacity [MW]	Status
1.	Kariba North Bank	Hydro	1080	Operational
2.	Kafue Gorge	Hydro	990	Operational
3.	Victoria Falls	Hydro	108	Operational
4.	Itezhi tezhi	Hydro	120	Partly commissioned
5.	Maamba	Coal Fired	300	To be ready in 2016
6.	Batoka Gorge	Hydro	1600	Pre-feasibility Studies done
7.	Kafue Lower	Hydro	750	Feasibility Studies done
	TOTAL:		4948	

The targets in Table 5.4, if achieved with good rainfall activity in the future, will be in harmony with the Vision 2030 which aims at producing 4337 MW of electricity by the year 2030. In fact, the current national energy demand is around 1600 MW and in 2030 it is unlikely to be more than 2500 MW. Thus, if the stations listed in Table 5.4 operate at almost full capacity, then load shedding would be eliminated and there would be enough power to export to neighboring countries like Zimbabwe, Botswana, Namibia and South Africa as was happening before.

However, studies (Yamba, 2011) reveal that seasonal climate variability will affect hydropower generation by reducing water run-off and reservoir storage capacity in dry years, and water overflow that can affect hydropower infrastructure in wet years. To mitigate such risks, they have recommended, the integration of renewable energy sources other than hydro in the power generation plans in order to sustain future demand. Such an approach will require undertaking a thorough study on the renewable energy potential (including both technical and economic) which has not yet been undertaken in Zambia. One way in which information regarding renewable energy potential can be obtained is through increased research activities.

Research activities would help find and develop solutions for adapting to climate change. The Government through the Ministry of Higher Education is encouraging topics on development and adaption of technologies for renewable energy and improving the performance and efficiency of renewable energy sources. For instance, The University of Zambia through the School of Engineering is currently carrying out one project entitled 'The Network of Energy Excellence for

Development (NEED) (Zorner, 2015). NEED sees Renewable Energy Technologies (RET) as a viable means of mitigating the energy shortages in Southern Africa. Five institutions are involved namely (i) The University of Zambia, looking at research strategies in RET, (ii) Namibia University of Science and Technology, looking at Dual studies and how its drylands of the Namib desert are to cope with complete reliance on RET as they are not connected to the national grid, (iii) Botswana International University of Science and Technology, is tasked to coordinate the development of the RET standards in the region, (iv) The Okavango Research Institute of Botswana is looking at how to adapt to RET implementation in the Okavango delta instead of the many diesel generators in the area, and (v) The Institute of New Energy Systems (InES) of Technische Hochschule Ingolstadt (THI) is the overall project coordinator and offers support in terms of management, and takes care of the network itself by ensuring knowledge transfer between the project partners and local stakeholders (Zorner, 2015).

Figure 5.7: NEED Fruitful Interactive Group Discussions during the Public Part of the Meeting with External Stakeholders in Botswana, September 2015 (Zorner, 2015).

When NEED partners meet, as shown in Figure 5.7, they always interact with the stakeholders to hear their views on the application of RET in their communities. When the project concludes in February 2017, a final report will be presented to Governments in respective countries.

Agricultural Solutions

The staple food for Zambia is maize. Feeding the whole nation requires highly mechanized (mechanical power technology) farming, which involves the use of energy. However, the majority of farmers in Zambia rely on human and animal draft power which negatively affects crop productivity. In responding to the drought of the 2015-2016 season, the Government, through its Disaster Mitigation and Management Unit, has launched the Disaster Management Policy and Settlement. This action will help in immediately identifying disaster-prone areas and quickly coming to the aid of the affected populace. The Meteorological department, for instance, has installed electronic rain gauges throughout the country and these send signals automatically to the data collection centre in Lusaka. The centre is able to help the Government in terms of identifying the areas that are likely to be affected with low farm productivity.

The Republican President has unveiled a three-option plan to deal with the looming food crisis in the wake of a devastating drought that has hit most parts of Zambia. The first is growing of irrigated maize to meet the shortfall. Irrigation would be sustained by creation of dams for water storage in farming areas of the country. The second option on the President's table is to engage the Food Reserve Agency (FRA) and farmers' union to "mop up" all the maize in the country, as well as to assess the deficit. The last option will be importation of maize if it is established that there is a deficit and local farmers cannot grow enough irrigated maize to fill the gap. That's why the Ministry of Higher Education encourages research on improving the productivity of communally grazed range lands and development of maize varieties that adapt to changing climate.

The Government has also embarked on the installation of solar-powered milling plants to reduce maize-meal prices and avoid dependence on grid electricity. A good number has been commissioned so far. To help farmers increase production of irrigated crop particularly in areas that are experiencing dry spells in the country, the power utility company has frozen the electricity tariffs for farmers. This is because crops require affordable electricity to support massive irrigation regimes for improved productivity to achieve food security for the nation while earning foreign exchange from exports of the surplus crop.

Conclusions

This paper has outlined a number of measures Zambia has taken to mitigate the effects of severe climate events. Due to reduced rainfall activity in recent years, the most affected are the energy and agriculture sectors. To combat the energy crisis, there are a number of projects embarked on such as bringing in a coal fired power plant for the first time in the history of Zambia, when the country has traditionally relied on hydro power generation. Thus, the country has found it important to invest more resources into water infrastructure and water harvesting because Zambia has numerous water bodies. To improve food security, the country is encouraging irrigation in times of severe droughts. This measure requires a sustainable supply of electrical energy to the agricultural machinery. Diversification into renewable energy technologies will help solve some of the energy problems in many households in the country.

Acknowledgements

The authors are grateful to the Centre for Science and Technology of the Non-Aligned and Other Developing Countries (NAM S&T Centre) for the financial support of the International Workshop in Colombo where this paper was accepted. Without this sponsorship, it wasn't just possible to prepare this paper for the Workshop. Thanks are also due to the NAM S&T focal point in Zambia for proposing the first author to NAM S&T Centre to have a return air-ticket bought for him to travel to Colombo to present the paper. Lastly but not the least, The University of Zambia is thanked for recognising the importance of such international events and supporting every effort that went into it.

Abbreviations

ZIPAR: Zambia Institute for Policy Analysis and Research

GRZ: Government of the Republic of Zambia

ACBF: African Capacity Building Foundation

MW: Megawatt

REA: Rural Electrification Authority

PSDMP: Power System Development Master Plan

M.O.E.: Ministry of Energy

ERB: Energy Regulation Board

CEC: Copperbelt Energy Corporation

DRC: Democratic Republic of Congo

MOU: Memorandum of Understanding

PV: Photovoltaic

RET: Renewable Energy Technology

NEED: Network of Energy Excellence for Development

InES: Institute of New Energy Systems

THI: Technische Hochschule Ingolstadt

FRA: Food Reserve Agency

NAM S&T: Centre for Science and Technology of the Non-Aligned and Other Developing Countries

ZEP: Zambia Energy Policy

WFP: World Food Programme

ESP: Energy Sector Profile

GDP: Gross Domestic Product

References

1. Energy Sector Profile, 2014. Zambia Development Agency.

2. Kuteya A., 2012. *Presentation at the Africa Lead Champions for Change Leadership Training*, Protea Hotel, Chisamba, Zambia.

3. Muzeya L, 2015. *Policy Design and Implementation in Developing Countries*, MEP14109, Ministry of Mines, Energy and Water Development, Zambia.

4. World Food Programme (WFP), 2015, "El Nino-Implications and Scenarios 2015," https://www.wfp.org/content/el-nino-implications-and-scenarios-2015-july-2015 (accessed 25 Feb. 2016)

5. Yamba F, *et al.*, 2011. Climate Change/variability implications on hydroelectricity generation in the Zambezi River Basin. Mitigation and Adaptation Strategies for Global Change, 16(6), pp. 617 - 628.

6. Zambia's Energy Policy (ZEP), 2008.

7. Zambia Institute for Policy Analysis and Research (ZIPAR), 2015. http://www.zipar.org.zm/ (accessed 22 Feb. 2016)

8. Zörner W, 2015.Network of Energy Excellence for Development, http://www.need-project.org/home.html (accessed 22 Feb. 2016)

Natural Disasters

Chapter 6

Natural Disasters and Climate Change in Nigeria: Issues and Mitigation Strategies

Agoro Olayiwola

Chief Scientific Officer,
Department of Physical and Life Sciences,
Federal Ministry of Science and Technology, Abuja, Nigeria
E-mail: layiagoro@hotmail.com, olayiwola.agoro@fedcs.gov.ng

ABSTRACT

Disasters whether man-made or natural are posing greater risks to the world natural ecosystem coupled with the ever increasing effects of climate change phenomena which had resulted to extreme weather conditions, droughts and pandemic outbreaks. The effects of all these disasters had resulted in biodiversity loss and depletion of genetic pool with little capacity for mitigation and adaptation. It is pertinent to state that Nigeria as a country is not immune to disaster of any type, and it had even happened severally in the country. Available statistics within a period of One hundred years showed that Nigeria had witnessed several disasters with the trajectory ranging from Drought, Epidemic, Extreme Temperature, Flood, Insects Infestation, Landslide, and Storm. Within the phase of these disasters, the worst was witnessed during the epidemic of 1991 with over 7, 300 lives lost, while during the 1983 drought related disaster, over 3m people were affected.

The incidences highlighted above manifested in destruction of properties and even loss of lives. Nigeria have had its own share of such incidents due to several reasons, one being the effects of climate change. One area that is increasing by the day is the menace of erosion owing to climate change effects. A place is question was the erosion which occurred at Nanka village in Anambra state where the Federal Government of Nigeria had spent millions of dollars at mitigating the effect of the disaster, yet no relief.

Keywords: Disasters, Climate change, Mitigation, Nanka and Federal Government of Nigeria.

Introduction

Nigeria is a country with a total area of 923,800 km² and occupies about 14 per cent of the land area in West Africa. The country lies between 4°N and 14°N, and between 3°E and 15°E. It is bordered respectively in the north, east and west by Niger, Cameroon and Benin Republic, while the Gulf of Guinea, an extension of the Atlantic Ocean, forms the southern border. The Nigeria area is covered with pre-cambian rocks, but there are also metamorphic and sedimentary rocks as well as volcanic rocks.

The highest areas are in the East, North and West where the land is generally over 1,500 metres, 600 metres and 300 metres respectively. The low lying areas which are generally below 300 metres lie along the coast and along the main river valleys. The Udi Plateau which lies to the east breaks the monotony of the surface along the coastal lowlands and these are characterised by coastal creeks and lagoons on both sides of the Niger Delta, while the main drainage system are the Niger-Benue, Chad and Coastal Rivers.

Nigeria is located within the tropics and therefore experiences high temperatures throughout the year. The mean temperature for the country is 27°C while the average maximum temperatures vary from 32°C along the coast to 41°C in the far north. The climate of the country varies from a very wet coastal area with annual rainfall less than 600mm with recent studies showing decline in rainfall. There are two seasons in the year; the wet and dry seasons.

The country has six main vegetation zones namely mangrove swamps, the fresh water swamps, the rain forest swamps and other coastal vegetation, tropical lowland rainforest, Guinea Savana, Sudan Savannah and the Sahel Savannah. Along the coast are salt water and freshwater swamps and other coastal vegetation. Further inland are the lowland rainforest, Guinea Savannah, Sudan Savannah and Sahel Savannah follow in that order.

The geographical location of Nigeria, its size and shape allows it to experience nearly all the different types of weather and climatic conditions of the West Africa sub-region. The vegetation varies regionally in consonance with the climatic pattern and its high population contribute to global warning. Ecologically, the country landscape encompasses the mangrove and fresh water swamp forest of the southern regions; the moist tropical lowland forest and savannah grasslands. The highlands vegetation of central Nigeria gradually transits to the scrublands of the distinctive semi-arid Sahel zone of northern Nigeria. The region has a fragile ecology which is exposed to the vagaries of climate change and the gradual encroachment by the Sahara desert to the north.

The Nigeria state has a population of over 180 million with very high population densities majorly found in the Eastern and Western States, Lagos Metropolitan Area and Kano State. In some parts of the Eastern states, the population exceeds 1000 person per sq km with the large areas of the country sparsely populated. There are over 250 ethnic groups, the numerous being Yoruba, Igbo, and Hausa/Fulani, while other groups include Tiv, Ibibio, Ijaw, Edo and Urhobo.

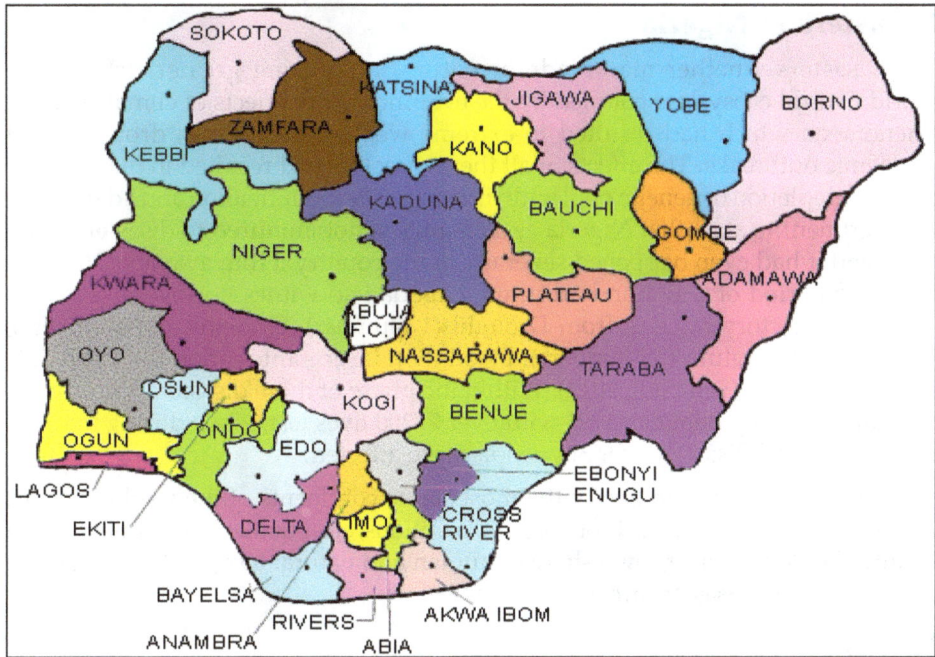

Figure 6.1: Map of Nigeria Showing the 36 States.

Disaster in the World

Disasters are hazards whether man-made or natural which poses greater risk to the world; both to the ecosystem and even to humans. The results of disaster most importantly in relation to climate change is ever increasing with greater effects on extreme weather conditions thereby resulting in drought and desertification, pandemic outbreaks, floods earthquake or landslide among others. Other deleterious impact of disasters in a natural habitat resulted in biodiversity loss and depletion of genetic pool with little capacity for mitigation and adaptation.

The occurrence and magnitude of natural and human induced disasters and emergencies are constantly becoming unpredictable and having greater consequences on our daily lives. Such disasters are posing greater risks to global natural ecosystem already rendered fragile by the culture of consumerism. Disasters in any where generally constitute the greatest threat to development including the socio-economic well-being of the people. It retards development and are very hard on the poor. According to a report by Action Aid (2006) based on the research by Adelekan Ibidun flooding is a disaster and was identified as a major factor which prevents Africa's growing population of city dwellers from escaping poverty and does not allow to improve life of urban slum dwellers. On the other hand, Climate change disaster caused by global warming remains a tropical issue globally with its wide range of impacts on vital sectors of the economy with considerable implications on humanitarian and developmental efforts.

Disasters in Nigeria

Disasters whether man-made or natural are posing greater risks to the world natural ecosystem coupled by the ever increasing effects of climate change phenomena which had resulted to extreme weather conditions, droughts and pandemic outbreaks. The effects of all these disasters had resulted in biodiversity loss and depletion of genetic pool with little capacity for mitigation and adaptation. It is pertinent to state that Nigeria as a country is not immured to disaster of any type, and it had even happened severally in the country. From available statistics in over a period of One hundred years, Nigeria had witnessed several disasters with the trajectory ranging from Drought (1 Number), Epidemic (50 Numbers), Extreme Temperature (2 Numbers), Flood (35), Insects Infestation (2), Landslide (3) and Storm (3) (Sani Sidi, 2012). Within the phase of these disasters, the worst it was during the epidemic of 1991 with over 7, 300 lives lost, while during the 1983 drought related disaster, over 3m people were affected.

Nigeria is located in one of the most stable continents of the world and this makes it less prone to hazards of volcanic eruptions, earthquakes and tsunamis. The country however, has its own share of human and natural hazards that constitute disasters. These disasters are as identified below:

Floods

The torrential tropical rainfall experienced annually across Nigeria is always associated with floods which had led to loss of lives, destruction of properties, infrastructure, shelters and farmlands among others. Other secondary hazards related to floods include outbreak of water borne diseases such as diarrhoea, dysentery, cholera and threats to food security. Nigeria as a country is very prone to flooding mainly along the Niger River through the Benue Basin and Sokoto Basin and this affects agricultural land. Many of the rivers in Nigeria equally have flood plains which are subject to flooding during the rainy season. Some of the rivers include; Rivers Niger, Benue, Cross River, Imo and Katsina. Urban floods also occurs in towns that are located on flat and low lying terrain most importantly those places where drainage has been blocked with waste, refuse and sediments. Such notable towns are Lagos, Ibadan, Aba, Calabar, Portharcourt.

According to report, about 25 million or 28 per cent of Nigerians live in the coastal zone and are at risk of flooding. In May, 2012 alone, 13 out of the 18 Local Council Areas of Cross River State were flooded with attendant loss of lives and properties recoded. The release of excess water from Ladgo Damlocated in the Northern Province of Cameroon on the Adamawa Plateau and takes it outflow through River Benue ravaged Adamawa, Taraba and Benue Stateswhen the dam attained the highest level in 29 years, this resulted to millions of people affected downstream communities in Niger, Kogi, Kwara, Kebbi, Anambra and Delta States. The floods led to land degradation and ecological dislocation. It is interesting to note that the sahelian states of Northern Nigeria which hitherto had been vulnerable to drought and desertification is now not exempted from floods as experienced in Sokoto-Kebbi in 2010 with the floods mainly nature induced and exacerbated by human activities mainly by the removal of plant cover. Construction of shelters within river basins, poor enforcement of town planning regulations and indiscriminate disposal of solid wastes along drainage channels are some of the causes of flooding. During the 2012 floods, a total of 7,705,398 persons were affected between July 1, - October 31, with 2,157,419 Internally Displaced Persons registered and over 363 confirmed dead. This indicates that flooding is a serious disaster that occurs where it was rarely recorded.

Figure 6.2: Submerged Houses Due to Floods in Numan in Adamawa State.

**Figure 6.3: Houses Submerged in Jigawa State,
North Eastern Nigeria Due to Floods.**

Figure 6.4: Post Effect of Floods with Several Properties Destroyed in Numan, Adamawa State.

The world is currently struggling on how best to pursue the four building blocks of mitigation, adaptation, technology and finance. It is important to contain adverse effects of climate change in Nigeria. In 2012 alone, flood disasters occurred in 23 states of the country including Sokoto, Kaduna, Katsina,Kebbi and Jigawa. With the event of 2012, more proactive effects was taken to prevent such again. The 2012 events gave the country the wake-up call that the country can no longer ignore disaster risks. The people at the grassroots are extremely vulnerable to cope with the effects of climate change with lack of assets such as land, livestock, income, education, healthcare, justice systems and conflict resolutions are responsible for their vulnerability.

Drought and Desertification

In 1972 – 73, there was a drought incident that affected the whole of Sahel zone of West Africa was very devastating consequence for the country. Some states in Nigeria such as Yobe, Borno, Jigawa, Katsina and Sokoto have continued to experience one form of drought which resulted in water shortage and poor nutrition. Desertification is a however a creeping disaster which has continued to claim several metres of arable land in northern Nigeria thereby giving rise to mass migration of people and animals southwards in search of opportunities. The rapid population growth, overgrazing, intensive cultivation and destruction of natural forest for biomass energy used in cooking and heating have accelerated biodiversity loss and depletion of genetic pool across Nigeria. The Sahara desert is creeping southwards with global warming due to human activities.

Ocean Surge and Marine Erosion

The extensive coastline of Nigeria has remained vulnerable to marine erosion and surge of tidal waters leading to destruction of properties, means of livelihood and loss of lives. Lagos, a commercial nerve centre of the country is under the threat of Atlantic Ocean. The reclamation of land had further aggravated the ocean surge with disaster steering if urgent action is not taken.

Landslide/Gully Erosion

Gully erosion has become a great disaster with very great consequence in Nigeria, this is prevalence South Eastern States of the country. The area is characterised by fragile sedimentary soil geology which is a big risk to several communities. In this area, land portion of lands had been taken over by gullies. This had lead to loss of lives and destruction of properties. It is moving at a very fast rate and our concern is that something urgent should be done on this.

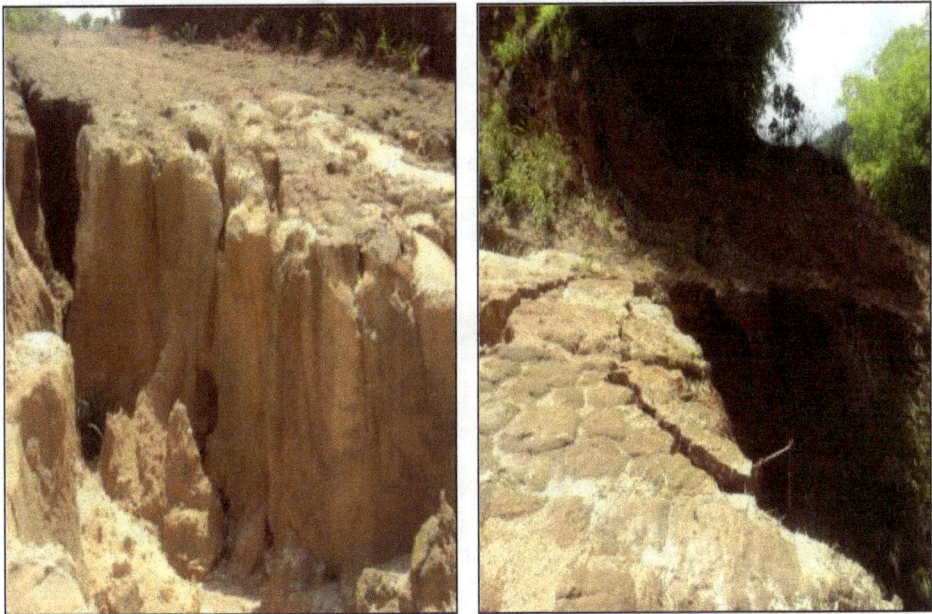

Figure 6.5: Landslide Leading to Gully Erosion in Nanka, Anambra State.

Land Use Conflict

This is one of the emerging sources of widespread disaster in Nigeria which if not quickly nipped in the bud will spell doom. Conflicts over dwindling land and water resources between farmers and pastoralists; two complimentary occupations have reached alarming proportions due to loss of lives and properties. This had happened in Benue – Nassarawa and Oyo States had all led to reduced agricultural production.

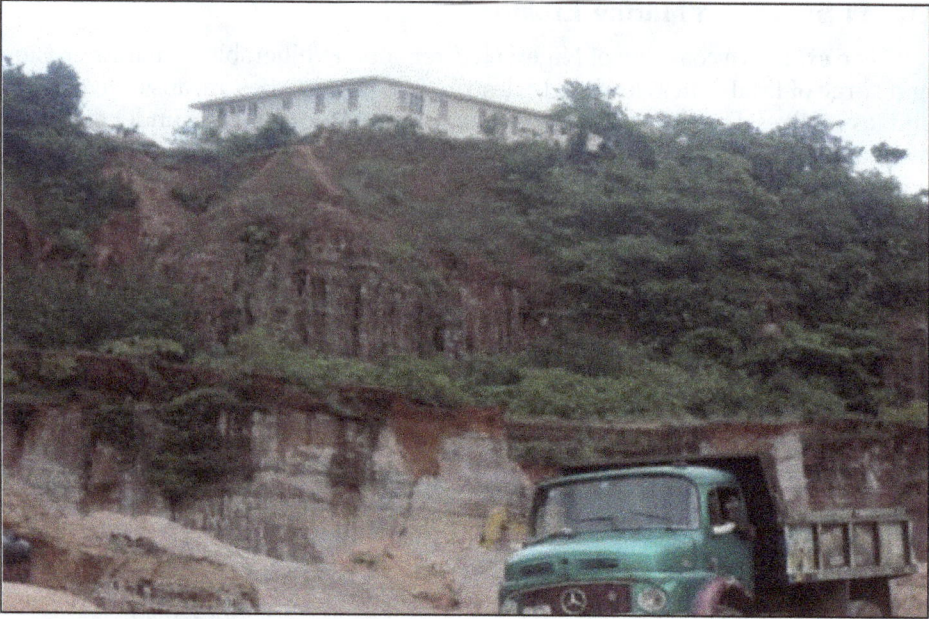

Figure 6.6: Landslide in Orlu, Imo State Showing a Threatened Student Hostel.

Figure 6.7: Landslide in Nanka.

Mining Related Disasters

One other area which Nigeria had received another share of disaster was that of sudden increase in infant mortality in Zamfara State which is suspected to be caused by lead poisoning as a result of crude mining activities in the rural communities. Over 400 infants (Children less than 5 years old) died and with over 500 people admitted in various hospitals across the states. The people were digging for gold while there was presence of lead in the same area and this led to water contamination in the affected area, hence poisoning.

Other Climate Induced Types of Disasters

Other Climate related natural disasters include flash floods, surges, cyclones, drought and severe storms among others have a negative effect on growth and the impact are considerable. It has thus become a reality with the inevitable occurrence. In 2010, the International Energy Agency (IEA) reported that greenhouse gas emission in 2010 were at their highest level in history, it is thus necessary for all to be up and doing most importantly developing countries who suffer from the impacts of climate change related hazards.

The UN equally reported that nine out of every ten disasters are now climate related with scientific evidences shown that with climate change, extreme weather events like floods, windstorm, droughts and epidemics have become more severe with severe of it occurring where they were unknown or extremely rare. It portray danger for us in the developing countries with the World Bank saying that losses from disasters can be up to 20 times greater as a percentage of gross domestic products in developing countries than the developed nations, while over 95 per cent of all disaster related deaths occur in developing countries.

Air disasters, communicable diseases and pandemic outbreaks. As pronounced in some countries in the world, incidences of natural disasters are increasing when compared with several decades ago. The incidences manifested in destruction of properties and even loss of lives. Nigeria have had its own share of such incidents due to several reasons, one being the effects of climate change. One area that is increasing by the day is the menace of erosion owing to climate change effects. A place is question was the erosion which occurred at Nanka village in Anambra state where the Federal Government of Nigeria had spent millions of dollars at mitigating the effect of the disaster, yet no relief.

According to report, Nigeria the level of vulnerability of the country to hydrometeorological hazards are high, and makes communities whose livelihoods depend on the natural ecosystem highly susceptible to climate change. The houses and farmlands of those communities located close flood prone areas are worst affected due to their lack of knowledge and awareness of the inherent hazards (NEMA 2012). The low level of income and lack of awareness to education, healthcare and economic base to cope with climate change was equally reported to exacerbate the effects coupled with the insanitary environment the people live.

The Nigeria National Capacity Assessment Report indicated that the country is prone to flooding mainly along the Niger River through River Benue and Sokoto Basin which affects agricultural land. Flooding in urban areas also occur in towns located on flat or low lying terrain (coastal areas) with little or no provision made for surface drainage and where existing drainages were blocked with municipal waste, refuse and eroded soil sediments. Several of the towns that had been

Figure 6.8: After Effect of the Landslide along the Enugu Expressway.

Figure 6.9: An Abandon Road under Construction Due to Erosion.

affected in the past included Lagos, Ibadan, Aba Calabar, PortHarcourt amongst others. Coastal erosion had virtually been experienced in almost all sections with the dire consequences shown by displacements of people and loss of lives as was experienced in Ogulaha community of Delta State in 2012. Over the years, there had been increase in the number of disaster incidents in Nigeria which are mostly hydro-meterological in nature and this had shown in negative impact of Climate Change in the country.

Challenges of Disaster Management in Nigeria

Disaster triggered by natural and human induced hazards in the country had claimed several lives. According to statistics by Odjugo and Oveyovwiroye (2012), the analysis of 10 natural disasters in Nigeria between 1900 and 2010 showed that the disasters are made of drought, flood and epidemic with drought affecting 3,000,000 people and flood leading to evacuation of 3,014,265 people and epidemic affected 80,000 people. Floods seem to be high stand this occurred between 1980 and 2000.

The report equally showed that the 10 deadliest natural disasters in Nigeria between 1900 and 2011 claimed 19537 lives and is epidemic in nature with bacterial infection (Cholera) affecting 10,000 and killed 7,289 people in two states of Bauchi and Kaduna. The ten costliest disasters in Nigeria destroyed property worth $189.5bn (N30.3trillion) between the same period apart from that of 1983 that destroyed property worth $71.7bn with all from flood.

National Institutional and Legal Framework to Combat Disaster in Nigeria

Due to the several incidences of disasters in the country, the Federal Government of Nigeria put in place several National Institutional and Legal Framework aimed at mitigating the effects of the disasters and most importantly climate change. These include:

1. National Science, Technology and Innovation Policy (2012)/Federal Ministry of Science and Technology
2. National Policy on Environment (1999)/Federal Ministry of Environment
3. National Guidelines and Standards for Environmental Pollution Control in Nigeria (1991)/National Environment Standards and Regulatory Agency
4. Waste Management Regulations/National Environment Standards and Regulatory Agency
5. Environmental Impact Assessment (EIA) Decree No. 86/Federal Ministry of Environment.
6. Natural Resources Conservation Action Plan/Federal Ministry of Environment
7. Nigeria First and Second National Communication/Federal Ministry of Environment
8. National Space Policy/National Space Research and Development Agency
9. Nigeria National Biodiversity Strategy and Action Plan/Federal Ministry of Environment

10. Nigeria is a signatory to the Hyogo Framework for Action (HFA) adopted in 2005 in Kobe, Japan provides roadmap for disaster risk reduction by building resilience of nations and communities of disaster losses

Funding Mechanism in Disaster Management

As a way of addressing the various kinds of Disasters in the country, the government had made provision for the funding mechanism as outlined below.

1. One per cent of the National Budget is allocated to mitigation of ecological problems and underlying factors
2. 20 per cent allocated to disasters mgt agency
3. Disaster Risk Reduction (DRR) funding by relevant agencies and department
4. States and Local Governments equally funds Disaster mitigation.

Strategies Adopted at Mitigating Disaster in Nigeria

The following Volunteers Teams were established and strengthened to assist in disaster:

a) Executive Volunteer against Disaster
b) Grassroots Emergency Volunteers

a and b are done in collaboration with local governments to train at least 200 people in their communities in emergency mgt skills.

c) Grassroots Emergency Vanguard,
d) NYSC Emergency Vanguard: National Youth Corp members are trained in disaster management techniques for the benefit of the host communities
e) Disaster Risk Reduction Club
f) Journalists Against Disaster

All these are capable of transforming emergency management landscape all over the country.

Steps take so far

☆ Flood vulnerability assessment studies undertaken for all the River Basin (12)
☆ Risk Assessment based on hazard data and vulnerability done.
☆ Vulnerability and capacity assessment done for Seven states of the country.
☆ Baseline studies on hazards and associated risks completed in Six states.
☆ Weather and climatic hazards monitored by Nigeria Meteorological Agency (NIMET).
☆ Damage and loss assessment methods and capacities done.
☆ Equipments to monitor seismic activities and eruption installed.

Figure 6.10: A New Seismological Station at the Federal University of Technology, Minna.

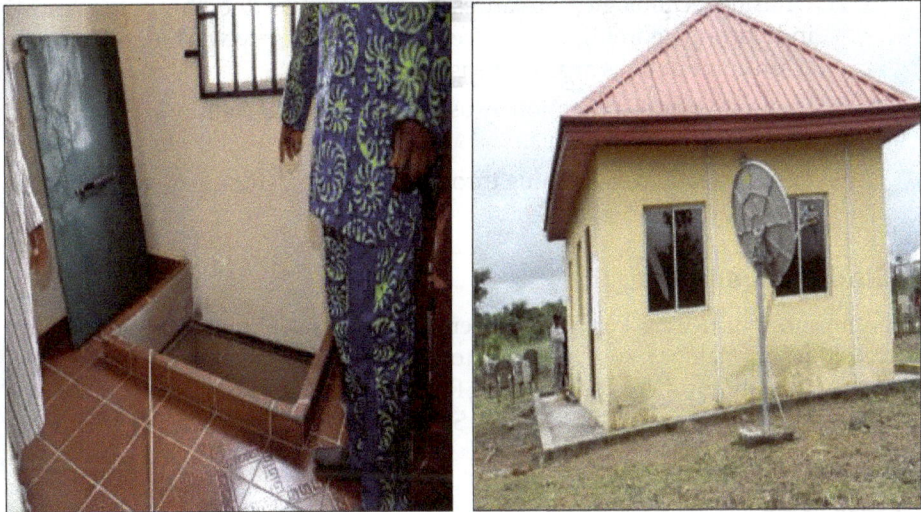

Figure 6.11: Another Newly Built Seismological Station at the Ebonyi State University, Abakaliki.

Way Forward

Concerted efforts are being put in place since the frequency and magnitude of natural and human induced disasters are constantly becoming unpredictable and having great consequences on present day survival. Such disasters are posing greater risks to global natural ecosystem already rendered fragile. This is further increased by global warming and the resultant climate change phenomena, droughts and pandemic outbreaks among others. Increased exploitation of natural resources has also increased biodiversity loss and depletion of the gene pool in developing countries with little adaptation measures. As such more should be done to move from the present policy to action such as:

1. Putting in place integration of Disaster Reduction into sustainable development policies, planning and programming at all levels with emphasis on prevention, mitigation, preparedness and vulnerability reduction.

2. Development and strengthening of institutions, mechanisms and capabilities at all levels in particular at the community levels to contribute to building resilience to hazards.

3. Incorporation of risk reduction approaches into the design and implementation of emergency preparedness, response and recovery programmes in the reconstruction of affected communities

4. Identify areas and monitor disaster risk and enhance early warning

5. Applying knowledge, innovation and education to build a culture of safety and resilience at all levels.

6. Sahara Green Wall Project to reduce desert encroachment.

7. Utilising technological advances in predicting weather through investing in hydro-meteorological services to establish early warning systems to warn communities against disasters

8. Using structural mitigation by using technological solutions such as flood levees and embankments.

9. Ensuring land use planning through adequate planning of infrastructure out of the harm's way.

Acknowledgements

I am grateful to the Permanent Secretary, Federal Ministry of Science and Technology, Dr. Habiba Lawal, the Director, Bioresources Technology Department, Pharm A. Oguntunde, Director, Planning Research and Policy Analysis, Mr. Babjide Oyelayo, Director-General, NEMA, Dr. Sani Sidi, My colleagues, Tope Fatogun, Chinedu Ikwunne, Peter Emmanuel.

References

1. Disaster Risk Reduction Capacity Assessment Report – 2012.

2. Ibidun O. Adelekan: Vulnerability of Poor Urban Coastal Communities to Climate Change in Lagos, Nigeria.

3. Impact of Natural Disasters on Developing Economies: Implications for the International Development and Disaster Community by Romulo Caballeros Otero and Ricardo Zapata Martzi.

4. International Journal of Mass Emergency and Disasters (March, 1995) Integrating City Planning and Emergency Preparedness. Some Reasons Why by Neil R. Britton and John Lindsay.

5. Muhammed Sani-Sidi (2013). Disaster Management is Everyone's Business, A collection of selected Speeches.

6. National Science, Technology and Innovation (ST and I) Policy (2012) by the Federal Ministry of Science and Technology, Abuja, Nigeria.

7. Nigeria's First National Communication (2003). Under the United Nations Framework Convention on Climate Change, November, 2003. Submitted to the United Nations.

8. Odjugo P. A. Ovuyovwiroye: Global Natural Disasters and their Implications on Human Sustainability.

9. The Informal Taskforce on Climate Change of the Inter-Agency Standing Committee and the International Strategy for Disaster Reduction.

10. Understanding the Economic and Financial Impacts of Natural Disasters by Charlotte Benson and Edward J. Clay.

Chapter 7

Case Study of Flash Flood Assessment of 14th November 2014 in Colombo due to Short Period High Intense Rainfall

A.R. Warnasooriya*, A.C.M. Rodrigo**
and K.H.M.S. Premalal***

Deputy Director,
Department of Meteorology,
Colombo, Sri Lanka
E-mail: *rashanthie@yahoo.com, **acmchanna@gmail.com,
***spremalal@yahoo.com

ABSTRACT

Flash flood in Colombo is becoming a frequent event with heavy and intense rainfall event, due to the land use change with rapid urbanization. Collapsed with the natural drain system with illegal constructions and less infiltration due to concrete and tar are some of the reasons which triggers flash flood event. In the international literature, it is possible to see many articles explaining the flash flood situation in several countries. Flash flood occurred in Colombo on 14thNovember 2014 at about 1630 caused due to heavy and intense rainfall. Many areas in the southwestern part of Sri Lanka received high intense rainfall and experienced flood situation. Coupled with heavy rainfall, strong winds also accompanied with heavy showers. Department of Meteorology, Colombo reported 56 mm rainfall within 20 minutes. Past flash floods in Colombo are always associated with heavy rainfall, but this event is not a common one due to less amount of rainfall. Analysis of pluviograph clearly evident the intense rainy condition. The aim of this study is to identify the cause of sudden changes of atmospheric phenomena during heavy precipitation on 14th November 2014. Observed meteorological data and data from NOAA NCEP was used in the analysis. Weather Research Forecast (WRF) model (Version 3.7.1) with 3DVAR data assimilation was performed.

In addition satellite images and atmospheric water content were also analyzed. 1.0 degree NCEP data were downscaled for 5 km grid to proper identify the local circulations. The results of analysis clearly showed that the change of local atmosphere with the reaching of middle level dry air mass combined with the moist atmosphere over Sri Lankan area. Recent heavy rainy condition in Sri Lanka was associated with the meeting of dry winds with moist winds (dry line). Therefore the findings of this study can be used for such events in future for development of early warning systems.

Keywords: Flash flood, Weather research forecast, NOAA NCEP, 3DVAR.

Introduction

Sri Lanka is an island located in the tropics between 05N to 10 N and 80E to 83E to the south of India. Sri Lanka is regulated by two (2) major monsoons as southwest (May – September) and Northeast (December to February) with the prevailing southwesterly and northeasterly winds respectively. There is another two monsoons named as first inter-monsoon (March to April) and second inter-monsoon (October to November) in between two major monsoons. 60 per cent of annual rainfall in the country is received from southwest and second inter-monsoon periods (Climatology available at the Department of Meteorology). 30 per cent of annual rainfall received during the second inter-monsoon and this is the highest rainfall intense period, because it is generally received within 2 months period and the rainfall is fairly distributed over the country. Heavy rainfall events within this period usually are associated with local thunderstorms as well as low pressure area developed in the Bay of Bengal.

Flash flood occurred in Colombo on 14th November 2014 at about 1630 SLST due to heavy and intense rainfall. Newspapers mentioned that, Gangarama, Armour Street and Norris Canal Road in Colombo impacted due to flash flood condition (Figure 7.1). Not only heavy rainfall, strong winds also accompanied with heavy showers. Similar condition was experienced even in Minuwangoda and Mattegoda areas according to the local newspapers.

Figure 7.1: Flash Flood in Colombo with Fallen some Trees on 14th November 2014 (Courtesy of Lankadeepa News Paper).

Rainfall observation at the Department of Meteorology, Colombo reported 56 mm rainfall within 20 minutes (Figure 7.2). Past flash floods in Colombo are associated with heavy rainfall, but this event is not a common one due to less amount of rainfall within a short duration. Analysis of pluviograph clearly evident the high intense rainy condition.

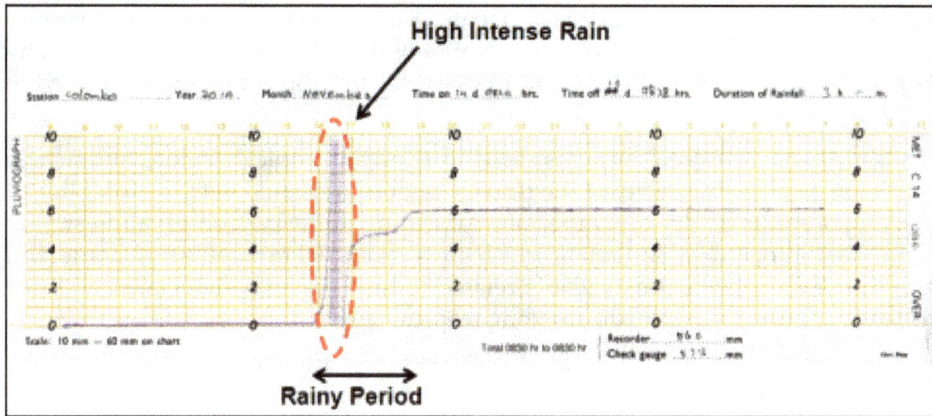

Figure 7.2: Pluviograph at CBO on 14thNov 2014.

Onset of southwest monsoon (end of May or beginning of June) associated with monsoon trough, development of mesoscale phenomena (thunderstorm) during inter-monsoon periods (March – April and October – November) and indirect impact of cyclones and low pressure systems developing in the Bay of Bengal, are some of the main reasons for heavy and intense precipitation in Sri Lanka. In addition, heavy rainfall can be associated with the Inter Tropical Convergence Zone (ITCZ), which is at the vicinity of Sri Lanka. Month of November belongs to the second inter-monsoon. Thunderstorm with heavy rainfall accompanied with strong winds can be expected during the second inter-monsoon. Even this is a general feature during inter-monsoon seasons, it can be triggered due to local changes of the atmosphere. There are many heavy rainfall cases in Sri Lanka, associated with mesoscale systems. Heavy rainfall and strong winds on 01stJune 2014 is a good example and 443.8 mm rainfall with heavy thunder activity with strong winds were received (Premalal *et al.,* 2015). Moore *et al., 2003;* Schumacher and Johnson 2005, 2006 mentioned that, a large percentage of extreme rainfall events result from particular organizations of deep convection in mesoscale convective systems (MCSs) that result in slow or repetitive storm motion over a particular geographic area. Even though the intense rainfall on 14thNovember 2014is associated with a thunderstorm, atmospheric phenomenon did not follow the usual pattern which is favorable for local thunderstorms, because it was found that dry mid tropospheric winds reached towards the western coast of Sri Lanka during the event.

It is well known that frontal systems are generally associated with strong convective mesoscale systems along subtropical region. Many studies have been conducted on heavy rainfall associated with frontal systems. Some of the examples are onset of the summer southwesterly monsoon over China region. It is generally occurred in the middle of May as the mei-yu front established its quasi-stationary position over southern China, Taiwan, and the western North Pacific (Tao and Chen 1987; Chen 1988, 1994; Chang and Chen 1995).On, 11 July 2007 indicates that a meso-alpha-scale low developed over a mei-yu front near 408N, 1348E at 0000 UTC (Biao, 2014).Themei-yufront extended from the East China Sea and swept across regions from western to eastern Japan. Deep convection evolved over western Japan well ahead of the mei-yu front. Not only these two events, there are many heavy and intense events can be found over the sub-tropical region associated to warm and cold fronts.

There are no frontal systems develop over the tropical region, but similar condition can be observed over the tropical region associated with the front line of meeting of midlevel dry and moist air. According to the past observation similar condition happened in Sri Lanka, which gave about 700 mm within 12 hours. But no written documents are available related to heavy rainfall associated with the collapsed of dry air and moist air along tropical region.

The aim of this study is to identify the synoptic situation associated with the heavy rainfall. Observation meteorological data and reanalysis data from NOAA NCEP was used for the analysis. Weather Research Forecast (WRF) model (Version 3.7.1) with 3DVAR data assimilation was used to downscale 1.0 degree NCEP data for 5 km grid to proper identification of local circulations. Satellite pictures and other atmospheric products also used to identify the clear reason for the intense rainy event on 14th November 2014.

Data and Methodology

NCEP (National Centre for Environmental Prediction) reanalysis data has been widely used for diagnostic studies as, they usually capture the synoptic scale meteorological patterns. In addition, the sub synoptic scale mechanisms responsible for the heavy rains were investigated by means of numerical simulations with a mesoscale model such as Weather Research and Forecasting (WRF) model. The WRF model is a mesoscale numerical weather prediction system designed for both atmospheric research and operational forecasting needs. It can generate atmospheric simulations using real data (observations, analyses) or idealized conditions. It features two dynamical cores, a data assimilation system, and a software architecture facilitating parallel computation and system extensibility. (Source -WRF home page). The WRF model version 3.7.1 was included in the study as a tool to understand the physical processes that cause the high intensity rains. The model offers advantage compared to re–analyses because it allows description of the dynamical processes with higher horizontal and vertical resolutions. The WRF model was configured with 3DVAR data assimilation with resolution of 15km for regional scale analysis and 05km for localscale horizontal grid size and 35 vertical layers. The numerical integration was carried out for 12hrs starting at 0000 UTC on 14th November 2014, using NCEP re-analyzes as initial and lateral boundary conditions. For the WRF model boundary conditions of topography are downloaded from the USGS website. Wyoming University sounding profiles were obtained to identify thermodynamic variables over the study region. Dundee satellite images were also downloaded for the study period to describe the sub-synoptic system.

In this study, the Runge-Kutta 3rd order time integration with 90 seconds time step were used in the WRF Model. WRF Single-Moment (WSM) 3-class and 6-class has been used as micro physics in the model and Kain-Fritsch (new Eta) scheme is being used as cumulas parameterization.

Model Configuration

The Grid Analysis and Display System (GrADS) software was used to visualize the relevant data such as Winds, Relative Humidity (RH) at the levels, surface, 850 hpa, 700 hpa, 500 hpa and 300 hpa. Wind shear, CAPE, Vorticity and vertical winds also visualized for better analysis to identify the reason. Mechanism for uplift low level moisture is one of the important factors to develop cloud. Favorable wind shear will be enhancing the uplift until to develop cumulonimbus cloud. Generally, low level wind shear is considered as one of the factors, to develop cumulonimbus cloud as Chaudhari *et al.*, 2010, explained. They have found that the favorable condition to develop longevity and strengthen thunderstorms in the tropical region is wind shear between 0.003 S^{-1} and 0.005 S^{-1}.

Table 7.1: Showing the Model Feature and its Configuration Used in this Study

Model Feature	Configuration
Horizontal spatial resolution	15km (Regional) and 5km (Local)
Grid points	100X80
Vertical Levels	35
Topography	USGS 30 sec
Dynamics	
Time Integration	Runge-Kutta 3rd order
Time steps	90 seconds
Physics	
Micro physic	WRF Single-Moment (WSM) 3-class and 6-class
Cumulus physic	Kain-Fritsch (new Eta) scheme
PBL	YSU scheme
Surface layer	Monin-Obukhov Similarity scheme
Radiation	LW- RRTM scheme, SW- Dudhia scheme
Land Surface Process	Noah Land Surface Model

Results

To verify the model performance, rainfall on 14th was compared with the observational values. Figure 7.3 shows the WRF (5km × 5km) model output and the observed rainfall and Figure 7.4 shows the model simulation with 15kmX15km resolution. Red dashed circle shows the rainfall over Colombo area and the condition is mostly similar. Therefore the model with both simulations were able to capture the inland rainfall to some extent, except some places. Although the both simulations able to capture the isolated fairly heavy fall, there were over estimation around black circled area in Figure 7.3(b).

Figure 7.5 shows the hourly wind pattern at 925 hpa to 500 hpa level from 1130 SLST on 14th November 2014. Only low level, upto 850 hpa, shows the development of localized wind disturbance over the southwestern part (red circled) of Sri Lanka, but the condition was gradually decreasing towards upperatmosphere. Therefore

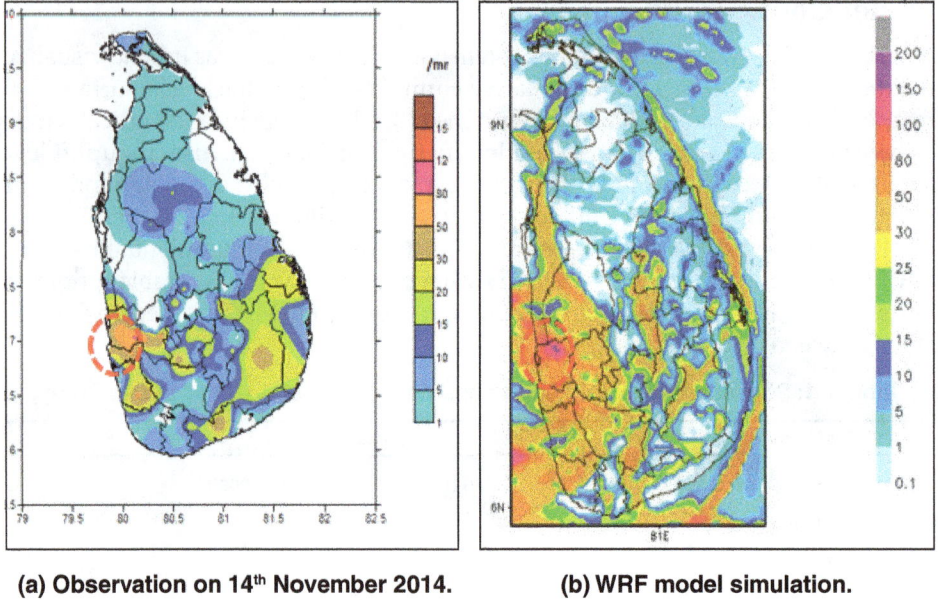

(a) Observation on 14th November 2014.   (b) WRF model simulation.

Figure 7.3: Observational Rainfall Data (a) and WRF Simulated Rainfall Data on 14th November 2014.

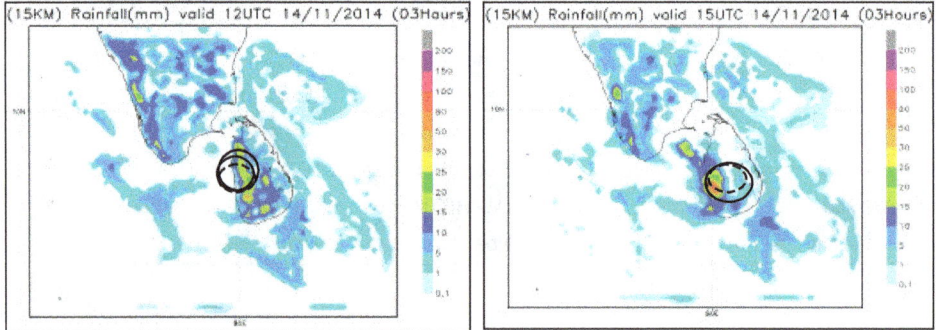

Figure 7.4: Hourly and 3 hourly Accumulated Rainfall (in mm) Distribution with WRFDA (15km × 15km) Model (with data assimilation) Simulated on 14th November 2014.

it clearly indicated the sudden change of loweratmosphere. Wind disturbance generally localized and it extended up to an 850 hpa level. Therefore it is clear that the intense rainfall is associated with a meso-scale phenomena.

It is essential to have moisture rich condition in the atmosphere to develop severe thunderstorm. Analysis of relative humidity indicated the dry atmospheric condition reaching towards the western coast at the middle level. Figure 7.6 shows the RH at 925 hpa (near surface) and 850 hpa at 1130 and 1730 SLST. No moisture

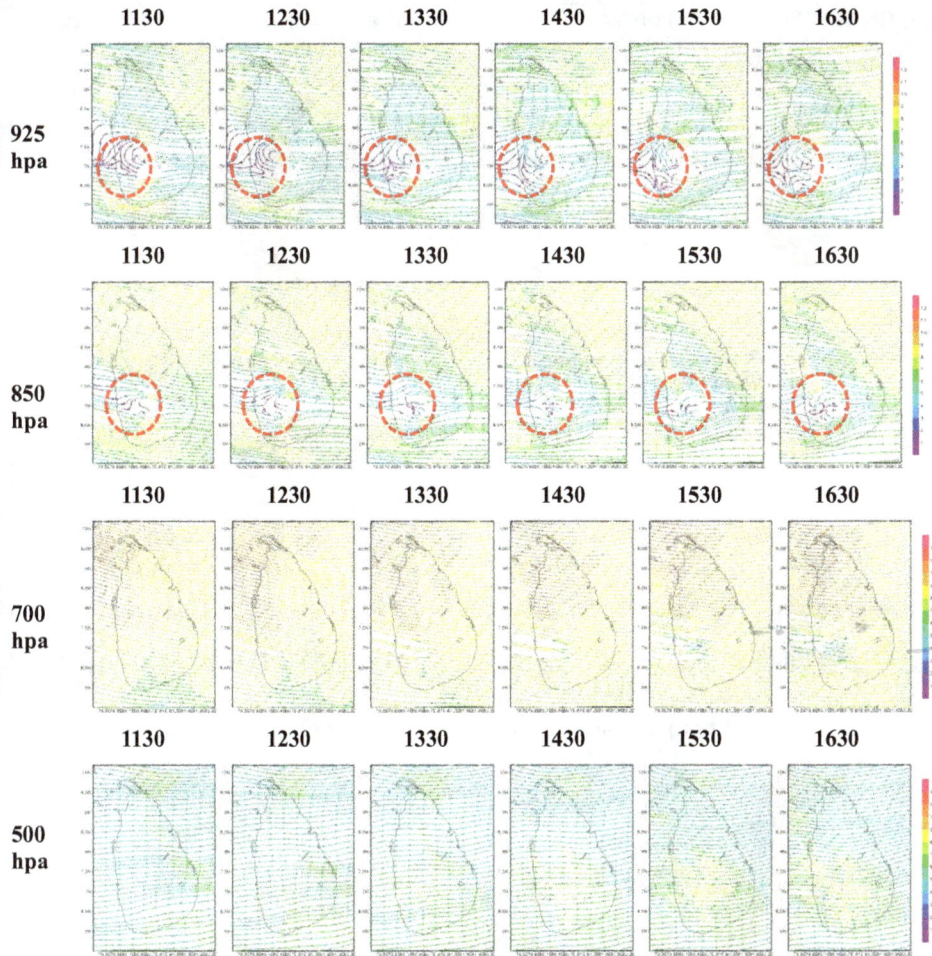

Figure 7.5: Change of Local Winds at 925 (Near surface), 850, 700 and 500 hpa Levels.

deficit at the surface, but it can be seen that gradual reaching of dry air parcel towards the western coast of Sri Lanka at the level 700 and 500 hpa (Figure 7.7). Satellite picture received at 1130 and 1730 SLST indicated the development of cloud associated to the meeting of dry and moist air (Figure 7.8). Local atmosphere was sudden change at the meeting edge of dry and moist air. Similar condition happened even in the past and heavy rainfall occurred due to the meeting of dry air and moist air. Therefore, the sudden change of local atmospheric and it upward motion was the reason of the meeting of dry and moist air. Wind shear was calculated for further clarification and it also indicated the sudden buildup of favorable wind shear (between 0.003 and 0.005 s^{-1}) for development of thunderstorm (Figure 7.8).

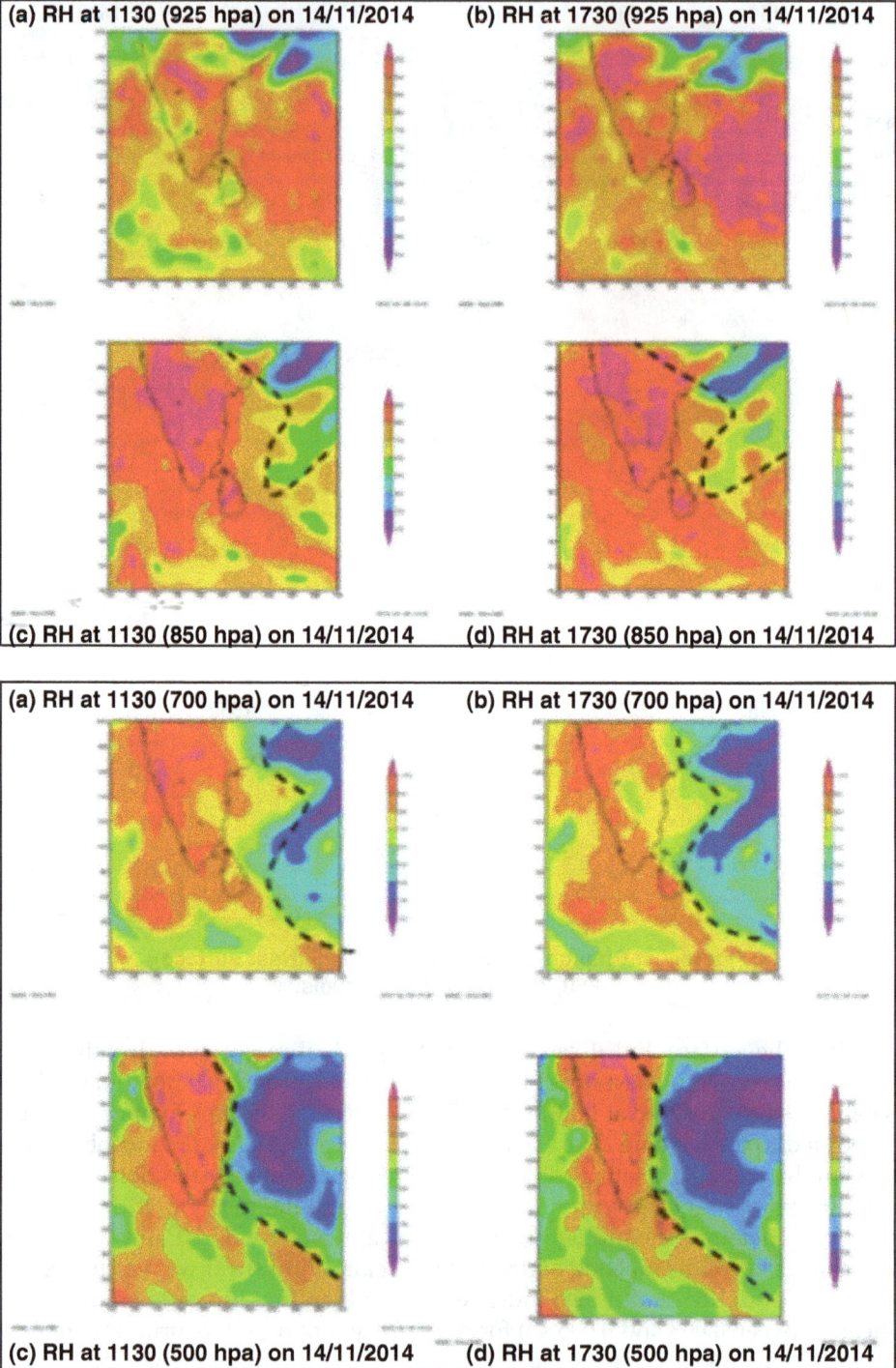

(a) RH at 1130 (925 hpa) on 14/11/2014

(b) RH at 1730 (925 hpa) on 14/11/2014

(c) RH at 1130 (850 hpa) on 14/11/2014

(d) RH at 1730 (850 hpa) on 14/11/2014

(a) RH at 1130 (700 hpa) on 14/11/2014

(b) RH at 1730 (700 hpa) on 14/11/2014

(c) RH at 1130 (500 hpa) on 14/11/2014

(d) RH at 1730 (500 hpa) on 14/11/2014

Figure 7.6: Relative Humidity at 925 and 850hpa (above), and
at 700 and 500bpa (below).

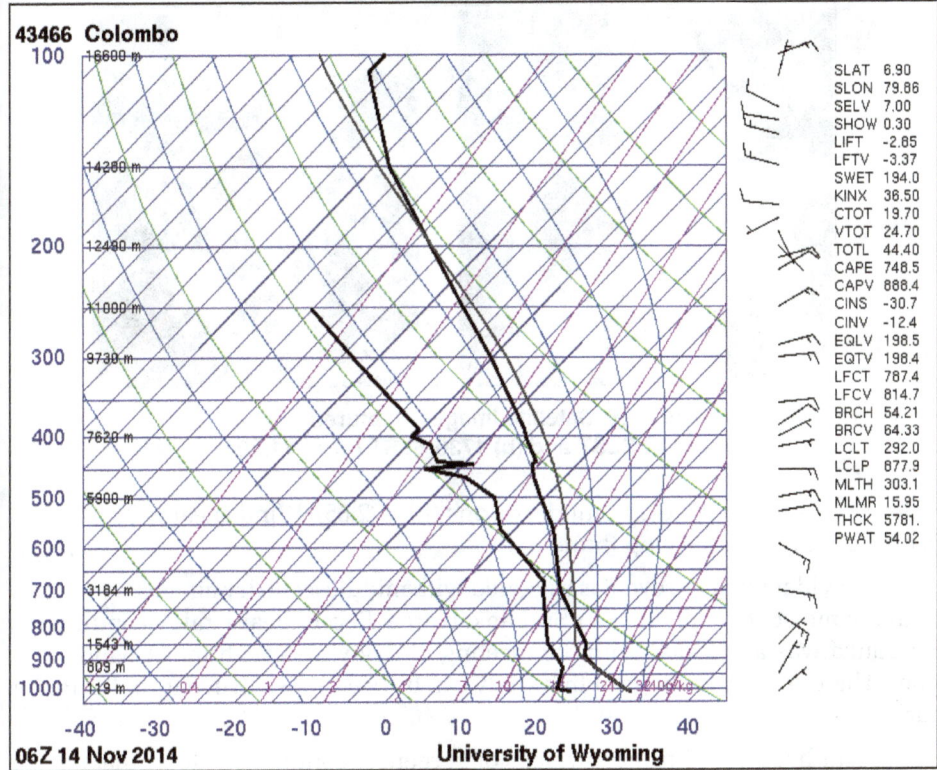

43466 Colombo

SLAT	6.90
SLON	79.86
SELV	7.00
SHOW	0.30
LIFT	-2.85
LFTV	-3.37
SWET	194.0
KINX	36.50
CTOT	19.70
VTOT	24.70
TOTL	44.40
CAPE	748.5
CAPV	888.4
CINS	-30.7
CINV	-12.4
EQLV	198.5
EQTV	198.4
LFCT	787.4
LFCV	814.7
BRCH	54.21
BRCV	64.33
LCLT	292.0
LCLP	877.9
MLTH	303.1
MLMR	15.95
THCK	5781.
PWAT	54.02

06Z 14 Nov 2014 — University of Wyoming

Figure 7.7a: SkewT Diagram at Colombo at 0600UTC on 14ᵗʰ November 2016.

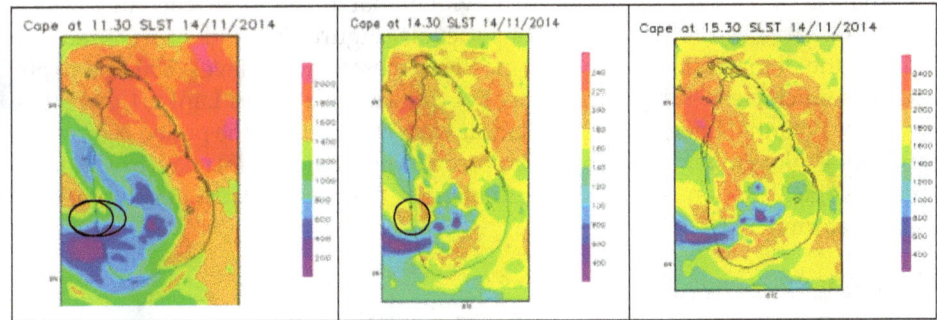

Figure 7.7b: Model Simulation of CAPE during 1130-1530 SLST.

Figure 7.7a shows the observed epigram at Colombo at 1130 SLST (0600 UTC) on 14ᵗʰ November 2014 and 7b shows the model simulation of CAPE during 11.30-1530 SLST on 14ᵗʰ November 2014. Model simulation is well matched with the observation at 1130SLST and shows the gradual increase of CAPE around Colombo area(Black circled) during 1430-1530SLST. At 0600UTC both show low CAPE (748.5), but Lifted

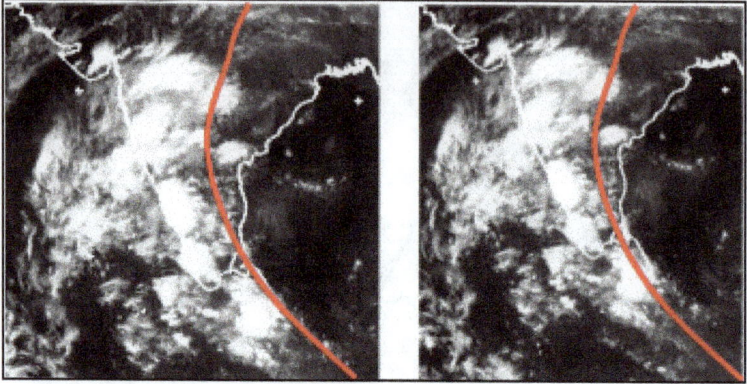

**Figure 7.8: Satellite Images Captured during
(a) 1130 SLST and (b) 1730 SLST on 14.11.204.**

index supported to develop thunderstorms (LI = -2.65). Wind shear is 0.007 and it is not supported to develop thunderstorm.

It is evident that the event is triggered with middle level dry air mass combined with the moist atmosphere over Sri Lankan area. Recent heavy rainy condition in Sri Lanka was associated with the meeting of dry winds with moist winds (dry line). Therefore the findings of this study can be used for such events in future for early warnings.

It can happen with the surface convergence, confluence etc at the surface level, but no convergence or confluence can be seen over the vicinity of Sri Lanka according to the wind analysis at different atmospheric levels shown in the Figure 7.4. Analysis of relative humidity clearly evident that dry air mass reaching towards Sri Lankan area at 700 hpa and 500 hpa levels (Figures 7.5 and 7.6). Dotted line indicate the demarcation of dry and moist air masses at the levels 850 hpa to 500 hpa. It is much clear that the movement of dry air towards Sri Lanka. Convective Available Potential Energy and Wind Shear were analyzed for the period 0000 – 1800 UTC on 14th November 2014. Upper air observation (Radiosonde) taken by Colombo meteorological station also analysed to identify the possibility of development of thunderstorm.

References

1. Biao Geng, 2014. Case Study of a Split Front and Associated Precipitation during the Mei-Yu Season, Weather and Forecasting, 29,996-1002.

2. Chen, G. T. J., 1988. On the synoptic-climatological characteristics of the East Asian mei-yu front (in Chinese with English abstract) *Atmos. Sci.*, 16, 435–446.

3. Chen, G. T. J., 1994. Large-scale circulations associated with the East Asian summer monsoon and the mei-yu over South China and Taiwan. *J. Meteor. Soc. Japan*, 72, 959–983.

4. Chang, C. P., and G. T. J. Chen, 1995. Tropical circulations associated with southwest monsoon onset and westerly surges over the South China Sea. *Mon. Wea. Rev.*, 123, 3254–3267.

5. Moore, J. T., F. H. Glass, C. E. Graves, S. M. Rochette, and M. J.Singer, 2003. The environment of warm-season elevated thunderstorms associated with heavy rainfall over the central United States. *Wea. Forecasting*, 18, 861–878,

6. Premalal, K.H.M.S., Warnasooriya, A.R., Rodrigo, A.C.M., 2015. Synoptic Analysis of Catastrophe Heavy Rain and Strong winds over Sri Lanka on 01ˢᵗ June 2014, 1, 50-63.

7. Schumacher, R. S. and R. H. Johnson, 2005. Organization and environmental properties of extreme-rain-producing mesoscale convective systems. *Mon. Wea. Rev.*, 133, 961–976.

8. Schumacher, R. S. and R. H. Johnson, 2006. Characteristics of U.S. extreme rain events during 1999–2003. *Wea. Forecasting*, 21, 69–85.

9. Tao, S. Y., and L. X. Chen, 1987. A review of recent research of the East Asian summer monsoon in China. Monsoon Meteorology, C. P. Chang and T. N. Krishnamurti, Eds., Oxford University Press, 60–92.

10. Weather Research Forecast (WRF) Home page.

Chapter 8

Children's Experiences of Participation in Disaster Risk Reduction: Perspectives from Parents and Stakeholders in Muzarabani

Chipo Muzenda-Mudavanhu

Lecturer,
Geography Department, Bindura University of Science Education,
Bindura, Zimbabwe
E-mail: chipomuzenda@gmail.com, cmudavanhu@buse.ac.zw

ABSTRACT

Children form the most photographed and vulnerable groups when disasters occur but are the least listened to members of the society. Little attention is being paid to building capacity of children to manage the impacts of disasters they experience now and those they will experience in future. This paper explores parents and stakeholders' views on children's participation and the contradictions that adults have to navigate in Muzarabani Zimbabwe. The study used the qualitative approach to document parents' (n=8) and stakeholders' (n=10) views about children's participation in disaster risk reduction (DRR). Results indicated that most parents do not always consult children during emergencies. Interestingly, was the belief among parents that children should be consulted when it is necessary in order to solve problems. The common dilemma among parents was the thought that it was wrong to put DRR responsibility on children and the fear of losing respect. On the other hand although, there was evidence of programmes where children were involved either in disaster risk awareness and response by different stakeholders most of the issues were on paper and not being practiced. The study therefore recommends the enhancement of DRR knowledge and skills at local level, creation of a school-community linkage, developing institutional capacity to deal with disasters and strengthening school physical conditions. This would facilitate the identification of children most at risk; facilitate their participation and address their particular needs in order to strengthen community resilience.

Keywords: *Children's participation, Disaster risk reduction, Experiences and community resilience.*

Introduction

Following the United Nations Convention on the Rights of the Child (UNCRC, 1989), four areas of children's rights were established, namely survival, development, protection and participation. The first three were said to be addressed by legislation but participation as stated in Article 12 of the UNCRC (1989) is often less supported in Zimbabwe. Article 12 (UN, 1989:4) states that;

> *"State parties shall assure to the child who is capable of forming his or her own views the right to express those views freely in all matters affecting the child, the views of the child being given due weight in accordance with the age and maturity of the child."*

The article allows considerable scope for interpretation of when children, individually and collectively, possess sufficiently mature capabilities to interact productively with adults but, Mitchell *et al.* (2008) state that the current research views children as passive victims with no role to play in communicating risks, participating in decision-making processes, or preventing disasters. There seems to be no model that singles out the potential role of children as resources in DRR (Twigg, 2004; Ronan and Johnston, 2005). Participation rights in disaster risk reduction (DRR) are viewed as aspirational and not yet fully realized (Alderson, 2008). Generally, interventions have been developed for adult populations and these have then been extrapolated for work with children. Thus, children's capacities to participate in DRR have been neglected (Haynes and Tanner, 2013). Yet children can act as informants within informal community communication networks, convey risk messages with a meaning and have a clear and uncluttered view about risks (SAARC, 2011). Of even more concern is the lack of empirical evidence to support children's participation in Zimbabwe. Therefore, the purpose of this study is to explore parents and stakeholders' views on children's participation and the contradictions that those engaged in children's participation have to navigate.

Children's Participation for Community Resilience

Despite the many challenges faced by children during disasters building children's ability to bounce back from a range of stressful situations is important in their long-term recovery from adversity (Mutch, 2014). To effectively and efficiently prevent, prepare and reduce disaster risks means finding ways of promoting children's participation in DRR. Children can change behaviours for more sustained development and are the leaders and decision makers of tomorrow (Plan, 2010). According Mutch (2014) to enhance resilience children should have a sense of safety and security, social connection, self-efficacy and sense of purpose, hope and meaning.

Although children's 'participation is not a replacement for adult responsibility, empowering children through their participation is an important protection strategy', as well as a right (Save the Children, 2006:2). Children's participation in DRR would ensure their safety. This is supported by Plan International (2010), arguing that participation and involvement in DRR fosters the agency of children to work towards making their lives safer and their communities more resilient. Because children are the most vulnerable group, during disasters, they need to be encouraged and motivated to participate in making the world a safer place to live. Izadkhah and Hosseini (2005:142) recounts that 'the theme of the United Nations 2000 World Disaster Reduction Campaign was Disaster Reduction, Education and Youth aiming at the continuation and development of a culture of prevention through education so that children can take a pro-active role in understanding risks and reducing the impact of disasters.' The campaign (2001) in Izadkhah and Hosseini (2005:142) stated that:

A culture of prevention is something that forms over time. What is needed is a change of attitude, based on the conviction that we do not need to be fatalistic about disaster risks and a willingness to act upon that conviction. The mind-set is best developed at any early age.

Thus, children's participation in DRR empowers them to make informed decisions concerning the risks of disasters. Though investing in child-focused DRR is a long-term process, it creates a generation that is better prepared for the disasters of tomorrow. Apart from empowerment of children through DRR, their involvement contributes to the realisation of their rights. The same was noted by Mitchell *et al.* (2008) that the approach recognises children as key actors in their own development and in their communities. Children's participation in DRR therefore is an entry point for programmes aiming at promoting sustainable development and children's rights.

Some immediate examples that can be utilised to enhance the participation of children in disasters include that children with illiterate parents can convey messages about DRR. They can also recognise disasters alongside social and economic threats (Mitchell *et al.*, 2008). Though they can convey disaster messages and recognise threats, they are usually not given the chance to do so. Given the resources, encouragement and the opportunity to participate, there is also a need to determine the manner in which children can build community resilience in their areas. The question that arises then is; what exactly are children expected to do in the society to build community resilience at the same time reducing their own risks?

While there is growing evidence that taking a child-centred approach in community based adaptation can build the adaptive capacity of children and also provide benefits to entire communities, Mitchell and Borchard (2014) note that there is no evidence to prove that what has worked in a growing number of cases is more broadly applicable, translatable to other regions or sustainable in the absence of direct project support. Children's participatory role is less supported than the engagement of older demographic cohorts because their role is understood differently among countries (Archard, 1993; Lister, 2007). Active involvement with children is least developed and most questioned because of its ability to undermine adult authority (Lundy, 2007).

Furthermore, reliable data on the actual number of children participating in DRR is limited with scientific journal articles rarely providing information on the magnitude of their involvement in DRR process. This could imply that much less has been published regarding children's capacities and contributions in DRR (Jabry, 2005; Peek, 2008). This is an important gap which this study responds to. There is need to increase our knowledge of children's capacities in order to better understand roles children play in vulnerability to disasters. This would provide evidence confirming the value of children's participation in DRR in order to facilitate their participation in marginal areas.

The Study Area

Muzarabani is one of the worst disaster prone districts in Zimbabwe in the Mashonaland Central Province, along the floodplains of the Zambezi River with Lake Kariba upstream and Cabora Basa downstream at the confluence of Musengezi and Zambezi rivers about 400m above sea level. Geographically it extends from 30° 45''E to 31° 20''E and 16° 00''S to 16° 30''S (Madamombe, 2004). It is found in the semi-arid and northern low-veld of Zimbabwe in agro-ecological zone IV characterised by little rainfall ranging between 450 to 600mm per year. Thus, the area is prone to seasonal droughts and severe dry spells in between the summer months (Campbell, 1994, Murwira *et al.*, 2012). Temperatures are excessively high (up to 40°C during the hot season – September to November) creating a suitable environment for mosquito breeding, hence the area is prone to malaria.

Trends in the occurrence of droughts indicate that they are becoming more frequent than ever before and are threatening Muzarabani District (Civil Protection Department, 2009). According to IPCC (2007) and Bola *et al.* (2014) floods and droughts are likely to get worse, as it is predicted that the magnitude and frequency of floods and droughts will increase during the 21st century due to changes associated with climate change and variability. Perennial flooding is the leading cause of losses from natural hazards and is responsible for a greater number of damaging events in Muzarabani.

Muzarabani is normally flooded from January to the end of the rainy season in March. The floods are mainly caused by localized heavy seasonal rainfall and run-off which often results in rivers overflowing. Chingombe *et al.* (2015) argue that the discharge from Mavhuradona range of mountains is drained and is propelled as a runoff wave that upon reaching the alluvial fans spreads out and fills the low lying area thereby leading to flash floods occurring. They further reported that the plain has a very small terrain variation hence bankful discharge is easily attained and the plain makes flooding of the area a very easy task for the river. Cyclones like those of February 2000 (Cyclone Eline) and March 2003 (Cyclone Japhet) are also responsible for flooding. The climate extremes (floods and droughts) can also contribute to the outbreak of diseases like cholera and malaria. Heavy rains tend to cause contamination of safe water contributing to the outbreak of cholera. The area forms the foot of and escarpment stretching from the east to the west of northern Zimbabwe.

There were 29 wards in Muzarabani District. Fourteen (14) wards were in the lower with 15 in the upper Muzarabani. Of the 14 wards three wards were identified as flood-prone areas which were; Chidodo, Hoya, and Chadereka. Other areas which are normally affected by flood disaster were not accessible as bridges were destroyed by floods and the roads were in poor state due to heavy rains experienced during the rainy season. Chadereka Ward 1 (Figure 8.1) was identified as the worst affected area (IFRCRCS, 2007) as it is at the confluence of Hoya and Nzoumvunda Rivers which are heavily silted. Compared to other wards Chadereka was found to be the most vulnerable community, and was selected for this study.

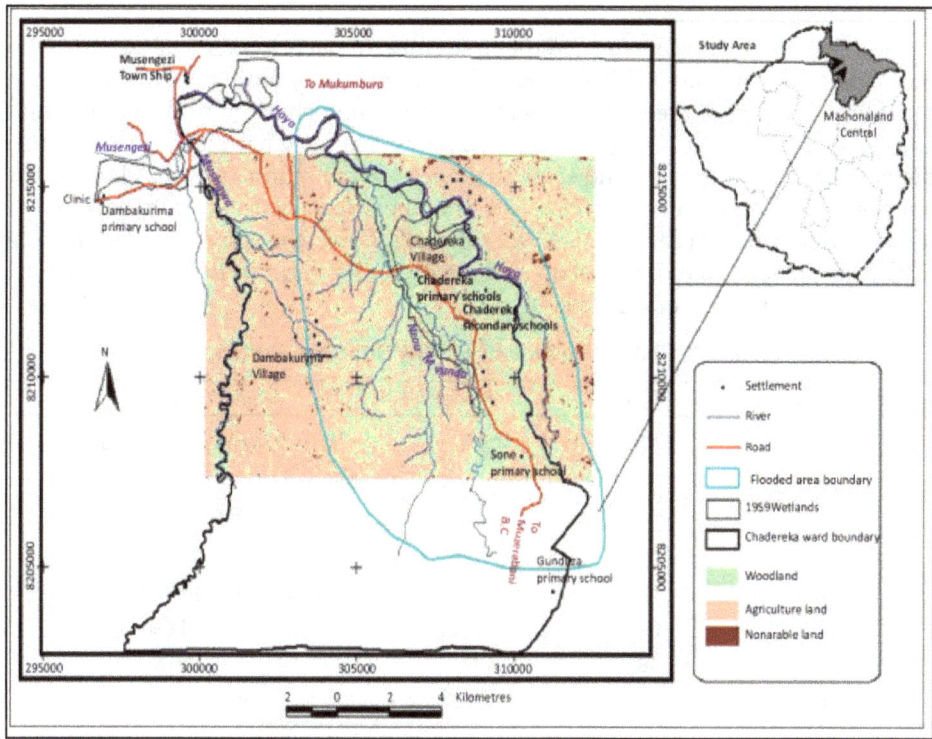

Figure 8.1: The Map of Chadereka Ward.

Methodology

This research adopted a qualitative case study approach. The purpose was to understand the issue of children's participation from the everyday knowledge and perceptions of specific groups of people. The researcher was interested in describing how things were experienced by those who were working with children.

Purposive sampling was used in the selection of the schools and also the participants (Punch, 2005). Four schools were identified in the normally flood zone and were considered to be the most vulnerable because of their proximity to the main streams in the area (Figure 1). The schools were highly recommended by several disaster practitioners who were working in Muzarabani.

There are however wide discrepancies in the recommendations in relation to sample size. For example, Fielding (1996) in Teye (2008) suggests 40 as the median sample size whereas Kvale (1996) recommends between 10 and 15. However, Morse (1991) advises researchers to "sample until repetition from multiple sources is obtained. This provides concurring and confirming data, and ensures saturation" (Morse, 1991:230). Saturation occurs "when gathering fresh data no longer sparks new theoretical insights" (Charmaz, 2006:113). Sandelowski (1986) argue that the sample size must be large enough to achieve data saturation and yet small enough to allow in-depth analysis.

As such, eight parents formed the first group of participants. The parents who participated in this study were school development committee (SDC) members. The SDC members were selected as they represented children in the community and parents at school. The parents had knowledge on what children face when they are in the community and their challenges in the school. This has helped the interviews to remain focused. Two SDC members were selected from each of the four selected schools (Table 8.1). At each of the schools, the researcher selected the SDC chairperson/vice chairperson and the secretary because they were the most active members in the committee.

Table 8.1: Characteristics of Parent/Guardian Participants

Interview Number	Age	Gender	Marital Status
1	47	M	Married living with spouse
2	52	F	Married, not living with spouse
3	53	F	Widow
4	61	M	Married, living with spouse
5	59	F	Married, not living with spouse
6	64	M	Married, living with spouse
7	48	M	Married, living with spouse
8	62	F	Widow

The interview themes included adults' understanding of children's participation, how they involve children and the common dilemmas. The interviews included the knowledge of children's rights to participate and to be consulted, policies/ strategies/plans in place to ensure that children's views were listened to. The interviews allowed greater flexibility within the structure and process through the use of open-ended questions to allow the respondents to express their thoughts on a subject freely. Though there was an interview guide, the researcher had to probe whenever it was required to do so. The interview times varied although most interviews took about one hour. Notes were taken in field note books, which were later transcribed.

Additionally, key informant interviews were conducted with officials responsible for coordinating child related disaster management in Muzarabani. Ten (10) key informant interviews were performed: four head/senior teachers, two officers of nongovernmental organizations; and one each with the ward councillor, a chief, a nursing sister employed by the Ministry of Health, and the assistant to the district administrator for Muzarabani (Table 8.2) Only those NGOs that were directly involved in the welfare of children and were working in Muzarabani during the time of study participated.

Table 8.2: Stakeholder Participants

Interview Number	Name of Institution	Gender	Position Held at the Institution
1	NGO 1	M	Field officer
2	NGO 2	M	Field officer
3	Nurse	F	Sister in Charge
4	Councillor	M	Ward Councillor
5	Chief	M	Chief
6	School A	F	Head teacher
7	School D	M	Senior teacher
8	Ministry of Local Government	M	Assistant DA
9	School B	M	Head teacher
10	School C	F	Deputy head teacher

Stakeholder interviews sought to describe the ways in which children were involved in DRR activities from an informed point of view and how they viewed children's right to participate. Questions asked included: 'Whenever you have DRR meetings do you include children?', 'Do you discuss disaster issues with children?', 'Do you consider children's views in planning any DRR activities?' The interviews also described the role of children in the society. Key informants were interviewed at their work places for their convenience and to have access to secondary data which was important for the study such as school registers, policy documents and hospital records. Each interview took an average of one hour.

Data collected was analysed manually where the researcher organised and read to capture the main themes that were emerging. The raw data was organised into word files and was reviewed, known as data cleaning. This was then followed by reading through all the texts to gain familiarity with the data and understand the main themes that emerged. Specific expressions and words were closely and carefully read. The transcripts were then coded by writing comments on the margin and were compiled into sub-themes. The topics were grouped according to their meanings and codes were assigned. Where a theme could not fit the existing codes, a new one was assigned. This was followed by the expansion of the subthemes into the broader themes. Finally the themes were interpreted in relation to the reviewed literature. The researcher maintained detailed records of data without interpretations. As such, the paper presents the most dominant views as quotations in the research and in some cases maintaining the language used.

Children's Participation Experiences

First, this section explores parents' views and perspectives on children's participation in DRR. It then assesses stakeholders' views and perspectives on children's participation in DRR. Finally it addresses the challenges that those engaged in children's participation have to navigate.

Adults' Views about Children's Participation

The aim of involving parents was to explore their views on the general practices of children's participation in DRR. Although parents acknowledged the labour provided by children, their frequency of consulting children was inadequate (Table 8.3).

Table 8.3 indicates that all of the parents do not always consult the children during emergencies. Interestingly, parents believed that children should be consulted when it is necessary in order to solve problems. Only one parent never consulted children, as they thought disaster issues were a matter between adults. They considered the issues to be too complex for their young children.

Table 8.3: Parents' Frequency of Consulting Children

Description of Parents' Frequency of Consultation	Frequency
Never, it is a matter between us adults	1
Rarely, when it is necessary in order to solve problems	5
Often, because I believe it is the best way to come to a solution to a problem	2
Always, the child has the right to know what is discussed about them	0
Total	8

However, although, parents acknowledged that they rarely consulted children, the main reasons for not consulting with them were not related to the rights of the child. Instead, these were related the community and family values. Common statements among participants included:

As parents we need to train our children so that they will be able to work for themselves and their children in future. We need to train them. There is also a lot of work that needs to be done in the home. Without the assistance of children we will not do much.

[Male, aged 47]

Adults acknowledged that most of the labour was provided by children because either the parents were busy with other household duties, as well as other community and family expectations. Some parents said they were too old to run around, for example, gathering wild fruits or herding cattle. Thus children were doing most of the household duties in response to a specific need. This was confirmed by Save the Children (2010) who reported that opportunities for children to participate are created not as a natural step to realise the right to participation but because of the passion of individuals or in response to a specific localised problem.

While most of the parents agreed that theoretically, the participation of children was important; their motivations to consult children differed (Table 8.4).

There were multiple responses with all respondents agreeing that children were taking part in household chores for family benefit (Table 8.4). None of the respondents reported that children were participating as part of government policy, rights of the child or as a demand from the children. This would mean that either parents were not aware of the policies or they were not giving children the chance to express their views. When asked about the policy that supports children's participation in DRR, all were not aware and agreed that they were not really giving children an opportunity to express their views even if they provide most of the labour in most household chores.

Table 8.4: Major Reasons for Rarely Involving Children in DRR

Reason	Frequency, n=8
As part of their rights	0
Family benefit	8
Government policy	0
Children demanding to be involved	0
Personal belief to involve everyone	7
Community values	6
Benefit the child	3

Parents' Dilemmas for Children's Participation in Disaster Risk Reduction

Parents faced several fears and dilemmas regarding child participation (Table 8.5).

Table 8.5: Parents' Dilemmas of Encouraging Children to Participate

Responses	Frequency
Children will lose respect for adults	3
Adults will lose control of outcomes	1
Children do not understand the consequences of their decisions	0
It is wrong to put DRR responsibility on children	4
Total	8

The common dilemma was the thought that it was wrong to put DRR responsibility on children and the fear of losing respect. One parent said adults would lose control of the outcome of decisions and considered that children do not understand the consequences of their decisions in DRR.

Furthermore, poor perceptions about the role of children in DRR were common among parents. Parents said it was be wrong to put DRR responsibility on children as they were not mature to make meaningful decisions. The parents could not trust the views from children. The same was reported by Protacio-de Castro *et al.* (2007) and Lopez *et al.* (2012). They questioned children's motivation and activities if done without parental guidance. This resulted in a parental conviction that they alone were responsible for giving orders and that the role of children was to receive and carry out duties instructed to them. Parents thought it was their duty to make sure that children's welfare is well catered for, as illustrated by the following extracts:

There's no need to seek children's views in obvious issues. We know what is good for our children and our decisions won't harm them but do them good. As a school council, we represent our children and want them to learn in a friendly environment.

[Female, 52]

Asking children what they want may mean that we have failed to lead them. We've seen that if children follow what their elders propose they'll be secure and do very well in life.

[Male, 47]

Parents were also afraid of depriving children's valued stage of growth as they engaged them into meaningful participation. Disaster experience was taken to be traumatic and could cause death or injuries among the children. Thus, allowing children to be involved in disaster preparedness, response and/or recovery would cause more harm than good. Several parents also illustrated examples of 10 to 14 year old children trying to cross overflowing rivers on their own with limited knowledge of the potential dangers. There were however, broader issues as parents argued that since they had evidence that children's minds were not fully developed, it would not be proper to give them a chance to express their views and take them into consideration. As a result, parents would rather decide where children should go and what they should do than ask for their opinions.

The other notable dilemma was the fear of losing control over the children when they become more confident and assertive. Children lacked life experience as they may come up with what they called "weird" ideas that parents fail to control such as not wanting to go to school and shouting at elders or teachers. As a result, parents preferred to have the children follow the adults' decisions. Fear of failure to control children's ideas and actions had negative implications for the society such as cultural erosion.

If we allow children to make decisions we may not be able to control them because children can't be trusted. Children need to be controlled that is why we sometimes beat them because even the Bible says foolishness is bound in the heart of a child, but the rod of correction shall drive it far from him. As a result, if we leave children to decide on their own, we may lose control.

[Female 53]

This was also supported by another parent saying:

Listening to children's views may bring chaos in our society. When we were growing up, our parents used to tell us what to do. I believe we should do the same with our children otherwise they will not do well in life. A time will come when they will decide on their own.'

[Male, 61]

Furthermore, children were not allowed to join in adult discussions and parents viewed children's participation as unimportant. Parents expected the children to have respect by doing whatever they were told to do without questioning. On the other hand, parents thought they should do whatever they want with and for their children. Listening to children's views was considered to be western oriented which could create social ills in the future, as illustrated in Box 1.

Parents thought that Western ideas would cause problems in their society as children may view themselves as equal with adults. As such parents were reluctant to honour up to children's participation.

Box 1: Parents' Views about the effects of children's participation on local culture

Pachivanhu pedu vana vanonzi ngavanyarare. Kwete nokuti havazivi asi vachiri vadiki kuti vazive zvavangada kuita. Ibasa redu kutungamirira vana (In our culture children are supposed to be led by the adults who know the best interest of the child).

[Female, 59]

We need to conserve our culture, this idea of giving children freedom will give us problems. In our culture, children are supposed to listen and not listened to. We cannot have a community being led by children. That is why we are having problems with our children. NGOs are talking about rights and protection of children, which they copy from Western countries. These are causing cultural erosion and because of that we are now experiencing problems such as rape, drug abuse among the children. As parents we cannot be told what to do by the children whom we are feeding and taking care of. We can share ideas but children should not take the lead or disregard what the elders say. Experience is the best teacher, we have experience and children should also experience something for them to stand up and voice their concerns.

[Male, 64]

Children are always children. They should be led by adults as long as the child is under our custodian. The idea of giving children a voice was never heard in our culture. Children can cause confusion and may suggest things that may not be accepted by the elders and our ancestors. Children should just be there to assist elders until they get their chance when old enough.

[Male, 48]

Source: Author

There was also a likelihood of children being manipulated by adults. Parents felt that children could be easily manipulated by people to complain against parents, teachers, other siblings and relatives and make inappropriate decisions. They noted that children could also be manipulated to be political activists on issues that they may not even understand. As a result the parents thought that children should not have a 'say' in DRR.

Stakeholders' Views about Children's Participation

This section goes further to assess stakeholders' views on children's participation in DRR in order to understand their perspectives. It was established that organisations employ different strategies when dealing with children. There was evidence of programmes where children were involved either in disaster risk awareness and disaster response. Children's participation in disaster risk awareness was evidenced by stakeholders' extracts presented in Box 2.

Box 2 illustrates that stakeholders were involving children in communicating risks to the community. This evidenced that children had the capacity to communicate effectively about risks to the wider community. Head-teachers also

---

Box 2: Children's Participation in Risk Awareness

We do awareness campaigns every year just before the rainy season in all the surrounding schools to make sure that children have the information about the disaster risks. Sometimes we give them pamphlets to share with their families, or we spend time talking to them about the challenges that could be faced such as outbreak of diseases and drowning. We also taught them what to do in case of an emergency.

[Assistant DA]

We get most our disaster information from the school children. It is easier for organisations to pass messages through the children because they get to every part of the community. I remember sometime we received cholera alerts from our children. They brought pamphlets and the chlorine tablets for water treatment.

[School C, teacher]

Every year before the rain season we announce to children to avoid crossing flooded rivers, herding cattle near rivers and streams and to be always home in case of heavy rains. We also teach them to report any incidences of sickness or missing persons.

[School A, Head-teacher]

We are in charge of children's health. Usually before the rain season we visit surrounding schools to alert them about the impending disasters such as cholera, malaria and floods. We normally arrange with the headmasters so that we can have a session of about 15 minutes with the children. We also encourage the children to take the messages to the community.

[Nurse]

indicated that they provided and raised awareness amongst pupils, for example, by running sessions on 'what floods are', 'what they do' and what can be done to reduce the impacts.

Furthermore, children participated in school rehabilitation and maintenance. The teachers revealed that children were the ones involved in cleaning the schools after the flood event. Children helped to fetch water and river sand for the reconstruction of a classroom block and the toilets. Stakeholders also indicated incidences where children participated in food for work programs that concentrated on gully reclamation and road maintenance, among other projects.

The other notable activity by the stakeholders for children was the provision of relief aid. The NGOs provided food, clothing and shelter for the families that were affected by disasters. Books and in some cases school uniforms were provided to children who lost almost all of their belongings. School based feeding schemes were also facilitated by the NGOs to cater for children who were coming to school on an empty stomach after the flood or during drought periods. However, these were relief aid and there was limited children's participation. Children provided labour in off-loading food stuffs and assisted their parents to carry the food home.

We do a lot of activities to help children affected by disasters. But we are more into disaster relief. As an institution, we normally come in when the president declares a disaster. We support children with food stuffs, school uniforms and books. In extreme cases we help in school rehabilitation. However, there is minimum participation of children because we will be trying to manage crisis.

[NGO 2, Field Officer]

However, stakeholders noted that most of the issues concerning children were on paper and were not being practiced. Although children's participation was important organisations viewed it to be:

☆ Time consuming as children spent time preparing for the activities and may take children away from school

☆ Costly as children cannot mobilise their own resources

☆ A new concept which lacks stakeholder participation

☆ Difficult to implement due to cultural differences

☆ Require political will

Promoting Children's Participation in Disaster Risk Reduction

Giving children an opportunity to express their views would ensure that their disaster related challenges are correctly addressed. This has the potential to make activities more sustainable, integrative and empowering as it becomes an increasingly integral approach in enhancing community resilience (Fernandez and Shaw, 2014). One very promising entry point for children's participation in DRR was through schools. Schools could play a vital role in every community (Shaw and Kobayashi, 2001; Stanley and Williams, 2000) and they need to be well secured and protected. Most schools in Zimbabwe are physically located at the centre of communities acting as a hub of knowledge, and in addition to education, have been providing public services, such as venues for church services and other community meetings. With the variety of services provided by schools it could be noted that they are a community hub that could build networks among people and organizations, especially when they have a common objective in reducing disaster impacts (Matsuura and Shaw, 2015). Thus, schools could play a crucial role in facilitating children's participation through: enhancing DRR knowledge, strengthening school capacity in dealing with disasters, school-community linkages, strengthening community capacity in supporting children's participation and external support.

This study advocates for what was noted by Domesian (1997) in Ozmen (2006:2) that;

To get people think in a preventive way, and to see the links between disasters, development and environment, one needs a mind-set that is best developed at an early age. A culture of prevention is something that forms over time. Cultural approaches and paradigms must be taught early and in schools to have real success.

However, although stakeholders agreed that most of the disaster knowledge from come through school children, analysis of the education curriculum indicated that DRR was not present in the formal education. The teaching of DRR was not regarded as the responsibility of the schools but of the Civil Protection Department (CPD) and other interested stakeholders. The CPD, UNDP and the Ministry of Education published a book on DRR for Secondary Schools but the books have not been taken up by schools. Furthermore, content analysis of the textbooks and the school timetable indicated limited and no space for DRR. This was also supported by the absence of DRR in the teacher training curriculum, given the strong commitment to DRR found in policy documents.

Besides the challenges of DRR education in Zimbabwe, school-based DRR education programmes were highlighted as having the potential to achieve great results in building a culture of safety. Participants highlighted the need for the Ministry of Education, Sport, Art and Culture to integrate risk reduction in preschool, primary, secondary and tertiary levels in different ways. Thus disaster education could be incorporated into cognitive development skills at preschool, environmental science at primary level and in selected subjects like Geography, General Science, and History among others at secondary level. At tertiary level stakeholders suggested the need for the provision of teacher training facilities, stand-alone courses, and infusion in different programmes. The teacher training programmes would help teachers to effectively teach DRR in their classrooms. This could help children to gain confidence, self-esteem, sense of belonging and responsibility as they participate in new experiences and learn new skills and information (Fernandez and Shaw, 2014). As children gain the DRR knowledge and skills they could change behaviours for more sustained development as future leaders and decision makers. (Plan International, 2010) and can relay the messages to the community. To raise risk perception Shiwaku *et al.* (2007) noted that more applicable information should be transferred to children and the community should be involved in school disaster education where children can take measures with them or do activities with them.

Consultations with the stakeholders indicated that all schools were built by the community with government assistance but due to rapid population growth, it was difficult to really consider the building standards but the aim was to promote the education for all campaign. The school buildings were generally poor causing interruption in education during floods. Reconstruction was also beyond school capacity due to lack of funds. However, to reduce children's vulnerabilities to floods, participants suggested the need to maintain infrastructure so that communities have access to important facilities and for schools to be able to prepare for, respond to and recover from disaster events. There is also need for external support so that schools can access funds to reconstruct schools and proper shelter for children. Stakeholders also highlighted the need for vulnerability and capacity assessment of the schools in order to come up with the activities for disaster preparedness, response and recovery. They highlighted the need to set up the community based support staff responsible for assessing the strength of school buildings, identify safe areas and evacuation routes in case of flooding and availability of contingency services.

However, DRR education and strengthening school infrastructure may not produce positive results in building community resilience to disasters. Children may have DRR knowledge but use of the acquired knowledge in the community calls for school-community linkages. Therefore, DRR activities should not be limited within the school premises; there is need to involve parents, local community and local government in the activities (Petal and Izadkhah, 2008; Matsuura and Shaw, 2014). Children's active engagement in community based action to disaster risks has benefits not only for the development of their own adaptive capacity, but it can also be a source of energy, resourcefulness and knowledge for broader community-based adaptation efforts (Bartlett, 2008; Tanner, 2010). As children interact with other children and adults, if they are well informed and supported, they can be effective channels of information, role models and agents of change (Mitchell and Borchard, 2014).

To this end, results indicate that there were links between the schools and the community in Muzarabani, which could promote children's participation. The communities were linked to the schools through the SDCs. Research indicated that all the schools had the SDC representing parents at school and children in the community. All the parents in the SDC had their children attending the same school they were representing. Apart from the link through the SDC, it was also noted that schools act as community centres where adults can meet discussing other community issues. Some of the classrooms were also used for church services as part of the school's community service. In addition, schools carry an important function as DRR hubs that become evacuation centres during emergencies (Matsuura and Shaw, 2014) and provide storage facilities for food aid. However, despite all these functions provided by the schools to the community, schools and communities are often disconnected (Matsuura and Shaw, 2014). In Muzarabani, schools and communities proved to be disconnected when it comes to DRR. Parents were doing their activities with limited participation of children and non-participation by school authorities yet children spent most of their awaken time with teachers. School personnel felt that they were being left out in community DRR activities yet the activities have the impact on the well-being of children. The school personnel also highlighted that sometimes parents decide to withdraw their children from school without consulting the school authorities. Teachers noted that sometimes parents withdrew their children due to lack of knowledge yet if they were working together they could find solutions to assist children to continue with their education even after a catastrophic event.

Further, this study indicates that the community could link with the school through informal DRR education where parents could be invited to the school or the community could invite children to community DRR meetings. Although respondents highlighted that children were disseminating DRR messages to the community, they felt that adults could also be invited to the school when they discuss or talk about disasters so that they will speak the same language with their children. However, one of the headmasters also noted that the school may do that but most parents were not supportive. It was commonly reported that parents were not attending even the SDC meetings where they talk about the welfare of their children at school. Parents were not very supportive when it came to school activities. The reasons for poor attendance were attributed to low levels of literacy among the parents and that some of the communities were too far from the schools for parents to travel to participate. But this could be resolved by motivating parents to be actively involved in school activities.

This study also highlighted the formation of cultural and performing arts where children could present disaster information to the community through music, dance, and theatre among others. Children could present flood related poems and music on special events such as rallies and other community functions where they can talk about disasters. Teachers noted that they could take advantage of school clubs like debate club, Christian Union that already existent in the schools and introduce disaster debate competitions where they could invite the community or as invited by the community. This would give children the opportunity to display their disaster knowledge to the community and their ability to assist the community in DRR. Thus children's playfulness and creativity could lead to the development of new methods or new approaches to assessing vulnerability and capacity to flood risks (Mitchell and Borchard, 2014).

Similarly, parent-teacher associations could also provide an opportunity for school-community linkage. Adults in the community had a great deal of local experience and knowledge of disaster where teachers could learn from. Communities have different copying mechanisms to survive in flood prone areas which should be passed on to the next generations. Apart from sharing with their children at home, parents could also share with the teachers who will then mix the traditional and scientific knowledge of coping with the children. Leaving the situation as it is may raise a confused generation where at home they believe in traditional mechanisms and at school they are taught the scientific methods and combining the two without proper guidance may not yield positive results. Hence, the school should fully utilize the local DRR knowledge by incorporating them into the school DRR education and in turn, disseminate the knowledge back to the community (Matsuura and Shaw, 2015) through parent-teacher interaction.

Teachers blamed parents as they were not forth coming when invited to the school. Schools had an open door policy where parents could come anytime for consultations but the turnout was worrisome. Parents were not attending even the end of school term consultations; yet these consultations were designed to facilitate the participation of parents in child education. But consultations were designed to facilitate the participation of parents in child education. Some parents were ignorant due to high levels of illiteracy in Muzarabani. Therefore, as stakeholders in DRR, teachers and parents need to collaborate to facilitate children's participation in DRR. When they share the same interests in solving disaster problems, they can find ways to bring school and outside-school communities together (Matsuura and Shaw, 2015).

On the other hand, one should note that, although schools play a significant role in strengthening community resilience, it is evident that they could face a number of challenges which requires; a policy change, development of tools and guidelines for safer construction, funding, development of DRR institutional arrangements within local authorities and support. Schools and communities cannot work alone; they need support from government and non-governmental organisations and the private sector to help support effective resilience building. In doing so, it is assumed that the school community would benefit from their knowledge and expertise in DRR, skills and manpower, additional funding sources and resources (Malalgoda and Amaratunga, 2015). The findings suggest that these stakeholders need to collaborate and coordinate the DRR activities in the schools. Support from adults is also important in order to make children's participation in DRR successful. According to the HFA progress assessment survey, children are gaining DRR knowledge and skills, but they do not have the supporting environment to put their knowledge and skills into practice (Fernandez and Shaw, 2014). Local authorities also need to undertake responsibility for ensuring that all new developments are constructed to appropriate standards such as the resilient building codes, disaster resilient planning, construction and maintenance guidelines, hazard and risk maps, set back zones and rural development plans, and can withstand a disaster event such as storms, cyclones and flooding. All these need to be developed to incorporate disaster resilient provisions into existing planning and building regulations. Mitchell *et al.* (2008) conclude by arguing that building a 'culture of safety' is a complex process involving different stakeholders, as children cannot take the responsibility alone. This leads to the conclusion that the implementation of DRR education programmes is a long process which should not be considered a stand-alone event, but should rather be repeatedly reinforced over time.

Conclusion

To strengthen children's disaster risk awareness, there is need to improve the way DRR information is disseminated. Since most DRR education programmes are conducted in schools, the development of the DRR curriculum would benefit children and their communities. There is also need to dedicate resources towards child-centered DRR to reduce the disaster impacts faced by children.

Although schools can play a significant role in strengthening community resilience, it is evident that they may face a number of challenges. The study therefore recommends for empowerment of the school community through policy change, development of tools and guidelines for safer construction, funding, development of a DRR institutional arrangements within local authorities and support. Schools and communities cannot work alone; they need support from government, NGOs and the private sector to help support effective resilience building through the provision of human and financial resources to strengthen their capacity to deal with floods. In doing so, it is assumed that the school community would benefit from their knowledge and expertise in DRR, skills and manpower, additional funding sources and resources (Malalgoda and Amaratunga, 2015). The findings suggest that these stakeholders need to collaborate and coordinate the DRR activities in the schools. Support from adults is also important in order to make children's participation in DRR successful. This was also noted during the Hyogo Framework for Action progress assessment survey that children are gaining DRR knowledge and skills, but they do not have the supporting environment to put their knowledge and skills into practice (Fernandez and Shaw, 2014). Local authorities also need to undertake responsibility of ensuring that all new developments are constructed to appropriate standards and can withstand storms, cyclones and flooding such as the resilient building codes, disaster resilient planning, construction and maintenance guidelines, hazard and risk maps, set back zones and rural development plans. All these need to be developed to incorporate disaster resilient provisions and it is important to mainstream DRR into existing planning and building regulations.

References

1. Alderson, P. (2008). *Young children's rights: Exploring beliefs, principles and practice.* Jessica Kingsley Publishers, London.

2. Archard, D. (1993). *Children: Rights and childhood.* London: Routledge.

3. Bartlett, S. (2008). 'After the Tsunami in Cooks Nagar : The Challenges of Participatory Rebuilding.'*Children Youth and Environments*, 18(1): 470–484.

4. Bola, G., Mabiza, C., Goldin, J., Kujinga, K., Nhapi, I., Makurira, H.and Mashauri, D. (2014)'Coping with droughts and floods: A Case study of Kanyemba, Mbire District, Zimbabwe.'*Physics and Chemistry of the Earth*, 67-69: 180–186.

5. Campbell, B.M. (1994). 'The Environmental Status of Save Catchment.'pp.21-24. In: Matiza, T. and Carter, S. A. (Eds). *Wetlands Ecology and priorieties for Conservation in Zimbabwe, Giland Switzerland:* International Union of Conservation of Nature and Natural Resources (IUCN).

6. Charmaz, K. (2006). *Constructing Grounded Theory: A Practical Guide Through Qualitative Analysis.* London, Sage Publications.

7. Chingombe, W., Pedzisai, E., Manatsa, D., Mukwada, G.and Taru, P (2015). A participatory approach in GIS data collection for flood risk management, Muzarabani District, Zimbabwe. *Arabian Journal of Geosciences,*8: 1029–1040.

8. Fernandez, G., and Shaw, R. (2014). Youth Council participation in disaster risk reduction in Infanta and Makati, Philippines: A policy review. *International Journal of Disaster Risk Science*, 4(3), 126–136.

9. Haynes, K., and Tanner, T. M. (2013). 'Empowering young people and strengthening resilience: youth-centred participatory video as a tool for climate change adaptation and disaster risk reduction.' *Children's Geographies*, 1–15.

10. IPCC (Intergovernmental Panel on Climate Change). (2007). *Impacts, Adaptation and Vulnerability. Contribution for Working Group II to the Fourth Assessment Report of the Intergovernmental panel of Climate Change.* M.L. Parry, O.F. Canziani, J.P. Palutikof, P.J. van der Linden and C.E. Hanson (eds). Cambridge: Cambridge University Press.

11. Izadkhah, Y.O., and Hosseini, M. (2005). Towards resilient communities in developingcountries through education of children for disaster preparedness. *International Journalof Emergency Management*, 3 (2), pp.138-148. Mahmood Hosseini, 2(3): 138–148.

12. Jabry, A. (ed.). (2005). After the cameras have gone: Children in disasters. London: Plan International.

13. Kvale, S. (1996). Interviews: An Introduction to Qualitative Research Interviewing, California: Sage Publications, Inc.

14. Lister, R. (2007). 'Why citizenship: Where, when and how citizenship?' *Theoretical Inquiries in Law*,8(2): 693–718.

15. Lopez, Y., Hayden, J., Cologon, K., and Hadley, F. (2012). Child participation and disaster risk reduction. *International Journal of Early Years Education*, 20(3): 300–308.

16. Lundy, L. (2007). 'Voice' is not enough: Conceptualising Article 12 of the United Nations Convention on the Rights of the Child.' *British Educational Research Journal*, 33(6): 927–942.

17. Madamombe, E. K. (2004). *Flood Management Practices -Selected Flood Prone Areas Zambezi Basin*, Harare: Zimbabwe National Water Authority.

18. Malalgoda, C., and Amaratunga, D. (2015). 'International Journal of Disaster Resilience in the Built Environment Article information.'*International Journal of Disaster Resilience in the Built Environment*, 6(1): 102–116.

19. Matsuura, S., and Shaw, R. (2015). Exploring the possibilities of school-based recovery andcommunity building in Toni District,Kamaishi. Natural Hazards (75): 613–633.

20. Mitchell, P., and Borchard, C. (2014). Mainstreaming children's vulnerabilities andcapacities into community-based adaptation to enhance impact. Climate andDevelopment, 6(4): 372–381.

21. Mitchell, T., Haynes, K., Hall, N., Choong, W., and Oven, K. (2008)'The Roles of Children and Youth in Communicating Disaster Risk.' *Children, Youth and the Environment*, 18(1): 254-279.

22. Morse, J. (1991). Qualitative Nursing Research A Contemporary Dialogue Sage, Newhury Park, Galifomia.

23. Murwira, A., Masocha, M., Gwitira, I., Shekede, M.D, Manatsa, D. and Mugandani, R. (2012). 'Vulnerability and Adaptation Assessment.'*Zimbabwe Second National Communication to the United Nations Framework Convention on Climate Change*, Harare, Zimbabwe, pp. 35-73.

24. Mutch, C. (2014). 'The role of schools in disaster preparedness, response and recovery : what can we learn from the literature ?Pastoral Care in Education.'*An International Journal of Personal, Social and Emotional Development*, 32(1): 5-22.

25. Ozmen, F. (2006). 'The level of preparedness of the schools for disasters from the aspect of the school principals. '*Disaster Prevention and Management, 15*(3): 383–395.

26. Peek, L. (2008). Children and Disasters: Understanding Vulnerability, Developing Capacities, and Promoting Resilience – An Introduction. *Children, Youth and the Environment*, 18(1): 1–29.

27. Petal, M., and Izadkhah, Y. (2008). Concept Note: Formal and Informal Education forDisaster Risk Reduction. Concept Note, (May). http: //www. pacificdisaster.net/pdnadmin/data/original/Formal_Informal_Educat_on_DRR.pdf. Accessed 15 May 2014

28. Plan International (2010). Child-Centred Building resilience Disaster Risk through participationReduction Lessons from Plan International. Plan, UK.

29. Protacio-de Castro, E.P., A.Z.V. Camacho, F.A.G. Balanon, M.G. Ong, and J.A. Yacat (2007). 'Walking the road together: Issues and challenges in facilitating children's participation in the Philippines.' *Children, Youth and Environments*, 17(1): 105-122.

30. Punch, K. (2005). Introduction to Social Research: Quantitative and Qualitative Approaches. London: Sage Publishers.

31. Ronan, K.R., Crellin, K, Johnston, D.M., Finnis, K., Paton, D., and Becker, J. (2008). 'Promoting child and family resilience to disasters: Effects, interventions, and prevention effectiveness. '*Children, Youth, and Environments,*18(1): 332–353

32. Sandelowski, M. (1986). 'The problem of rigour in qualitative research.' *Advances in Nursing Science*, 8: 27-37.

33. Save the Children (2006). *75 years of lasting change for children in need.* Annual Report, Save the Children.

34. Save the Children. (2010). 'Regional study of children ' s participation in Southern Africa : South Africa, Swaziland and Zambia.' *Research Undertaken By iMEDIATE Development Communications in Association with ON PAR Development*, (March), 1–75.

35. Shaw, R., and Kobayashi, M. (2001). *Role of schools in creating earthquake-safer environment disaster management and educational facilities*, UNCRD, Nagoya.

36. Shiwaku, K., Shaw, R., Kandel, R. C., Shrestha, S. N., and Dixit, A. M. (2007). Futureperspective of school disaster education in Nepal. *Disaster Prevention and Management* 16(4): 576-587.

37. South Asian Association for Regional Cooperation (SAARC). (2011). Framework for Care, Protection and Participation of Children in Disasters. UNICEF and Save the Children.

38. Stanley, P., and Williams, S. (2000). *After disaster: Responding to the psychological consequences of disasters for children and young people.* New Zealand Council for Educational Research, Wellington.

39. Tanner, T. (2010). 'Shifting the Narrative: Child-led Responses to Climate Change and Disasters in El Salvador and the Philippines. *Children and Society,* 24(4): 339–351.

40. Teye, J. (2008). *Forest Resource Management in Ghana: an analysis of policy and institutions.* PhD Thesis, University of Leeds.

41. Twigg, J. (2004). Disaster Risk Reduction: Mitigation and Preparedness in Development and Emergency Planning. London: Overseas Development Institute.

42. UN (United Nations). (1989). *United Nations convention on the rights of the child.* Geneva: United Nations.

Disaster Management:
Practices and Policies

Chapter 9

Disaster Management Practices in Cambodia

Viseth Ung[1] and Leng Heng An[2]

[1]Deputy Secretary General,
National Science and Technology Council,
Ministry of Planning, Phnom Penh, Cambodia
E-mail: viseth_ung@yahoo.com
[2]Assistant to Secretary General,
National Committee for Disaster Management (NCDM),
Royal Government of Cambodia and Visiting Researcher of Asian Disaster Reduction
(ADRC VR 2013 B), Cambodia

ABSTRACT

Around the world, climate change and disaster has become the focus of urgent discussion and action. It presents a multi-dimensional challenge, but can also be seen as providing opportunities. Dealing with climate change marks a new paradigm for development. The Royal Government of Cambodia considers disaster management as a key component of its social and economic planning. Flood and droughts have caused serious damage and loss to Cambodia, and endanger the Royal Government's efforts to enhance the economy and well-being of Cambodian society.

National Committee for Disaster Management (NCDM) is established for overall coordination for disaster management. Headed by the Prime Minister, NCDM was set up in 1995 that consists of 22 members from different Ministries, Cambodian Armed Forces, and Civil Aviation Authority as well as representatives of Cambodian Red Cross. As part of the decentralization process, disaster management institutions have been set up at Sub-National level. National Strategic Development Plan (NSDP) 2014-2018 and the Strategic National Action Plan on Disaster Risk Reduction 2008-2013 work as the overarching frameworks, and provide strategic direction to disaster risk management for the country. Meanwhile, this objective of this paper is to show the information and results findings in disaster management practices in Cambodia for policy-makers, academia, researchers, students, interested stakeholders and relevant. This paper has been re-compiled by the Mr. Viseth Ung and yet still accredited and endorsed for Mr. Leng Heng An, Assistant to Secretary General, National Committee for Disaster Management (NCDM), Royal Government of Cambodia, for the first own report and during his term as a visiting Researcher of Asian Disaster Reduction (ADRC VR 2013 B) at NCDM.

The main findings of this report shows that in 2011, floods affected 350,000 households and 52,000 households were evacuated, 250 people lives were dead and 925 kilometers of the national, provincial and urban roads were affected and 360 kilometers of the roads were damaged. The 2011 floods caused an estimated loss at 630 million USD. Drought severely affects farming productivity especially among rice growing communities who rely solely on rain or river-fed irrigation. In 2012, drought hit 11 out of 24 provinces, affected 14,190 hectare of rice fields and destroyed 3151 hectares. The country has achieved remarkable knowledge and skills to live with disaster risk. Cambodia also endorsed the Hyogo Framework for Action in 2005 that provides a systematic and strategic approach to reduction of vulnerability and risk to disasters.

All in all, the finance will represent a significant addition to current official development assistance (ODA), and as such, an opportunity to overcome Cambodia's underlying vulnerabilities to climate change and disaster risk reduction while also creating conditions for investment in more long-term, low-carbon development pathways. Given the long-term nature of building climate change resilience and disaster management, it is essential that the availability of finance is secure to allow for the kinds of long-term planning and investments required.

General Information about Cambodia

Geography, Topography and Land

Kingdom of Cambodia (Cambodia) is geographically situated in Southeast Asian region and bordered with Vietnam to the east, Thailand to the west, and

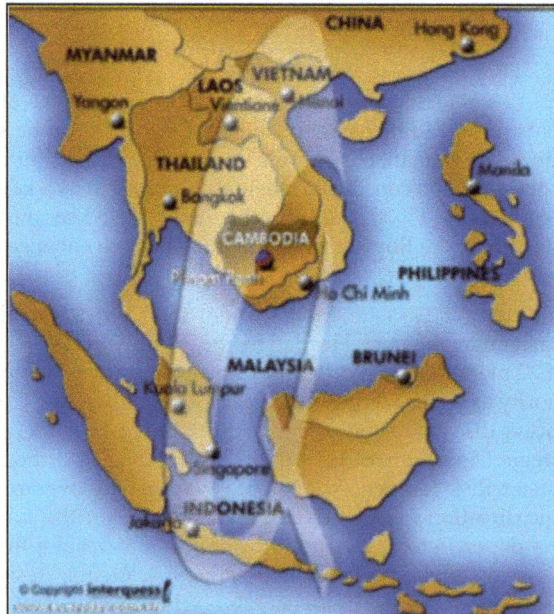

Lao PDR to the north and Gulf of Thailand to the south (Latitude: 10°- 15°N, Longitude: 102° - 108°E). Topographically, Cambodia is deep and plain at the middle, surrounded by mountain and Plateaus in the southwest lies the coastal area.

Cambodia has the land area of 181,035 square kilometers. 500 kilometers of Mekong River bisects Cambodia, and roughly 80 per cent of Cambodia's land is in lower Mekong basin, which their livelihood depend largely on agricultural production. Tonle Sap Lake, one of the world largest lakes, is situated in the mid-west of the country.

Climate

Tropically humid, the climate is dominated by monsoon with two seasons: rainy season from May to October and dry season from November to April. The average temperature is from 21 to 36 °C. The months with the lowest temperature are December and January while the months with the highest temperatures April and May.

Population

Cambodia has the population about 15 million (NIS, 2013), comprised of Khmer 90 per cent and other 10 per cent including ethnic minorities: Cham, Chinese and Vietnamese and indigenous and mountainous tribes.

Religion

Theravada Buddhism is the official religion in Cambodia which is practiced by 95 per cent of the population-- just like that of Thailand, Burma, Sri Lanka, and the rest follows Islam, Christianity and animism.

Government

Cambodia is Constitutional Monarchy and has adopted Parliamentary Representative Democracy. The Prime Minister is elected in majority vote among 123 members of the National Assembly, who are elected every five years in the general election and officially appointed by the king. The Prime Minister of Cambodia is the head of government while the King is the head of state. The current king is His Majesty King Norodom Sihamoni, and the current Prime Minister is H.E Samdech Hun Sen.

Administrative Division in Cambodia

Administratively, the country is divided into 25 provinces and one capital city, District Level: **197** districts/*khans*/towns, and **1633** communes/sangkats. Phnom Penh, the capital city of Cambodia is the political, socio-economic and cultural center of Cambodia.

Economy

Cambodian currency is KHMER Riel (1 USD~ 4000 Riels). Cambodian economy largely relies on agriculture production (rice, rubber, fish, corn, wood, vegetables, cashew nuts...), tourism sector, light industry (textiles and shoes production) and construction.

Natural Hazards in Cambodia

Natural Hazards Likely to Affect the Country

Disasters are inseparable from economic, social and environmental features of Cambodia. The country experiences almost all types of hydro-meteorological hazards such as floods, drought, heavy storms (or typhoon), fire incidents and epidemics. Most geographical regions of the country (*i.e.* Riverine Central Plains, coastal ecosystems and Dangrek mountain range in the north and Cardamom mountains in the southwest) are exposed to one or more of these hazards. Additionally, climate change is expected to increase the frequency, intensity and severity of these extreme natural events. As the majority of Cambodians are farmers and their livelihoods mainly depend upon subsistence agriculture, the vulnerability of people living in rural areas is very high and may continue to rise, requiring improved preparedness and planning.

The mighty Mekong River that enters the country from Laos and Great Tonle Sap Lake in the middle created the unique flooding feature, and most typhoons originate from the South China Sea towards the south and southeast across Vietnam and Southern China.

Floods

Cambodia is one of the five countries located along the Mekong River, and its landscape consists of rolling plains and lowland. During the monsoon season, Cambodia experiences flash floods usually after heavy rainfall. The provinces of Battambang, Kampong Chnang, Kampong Speu, Kampong Thom, Kampot, Kandal,

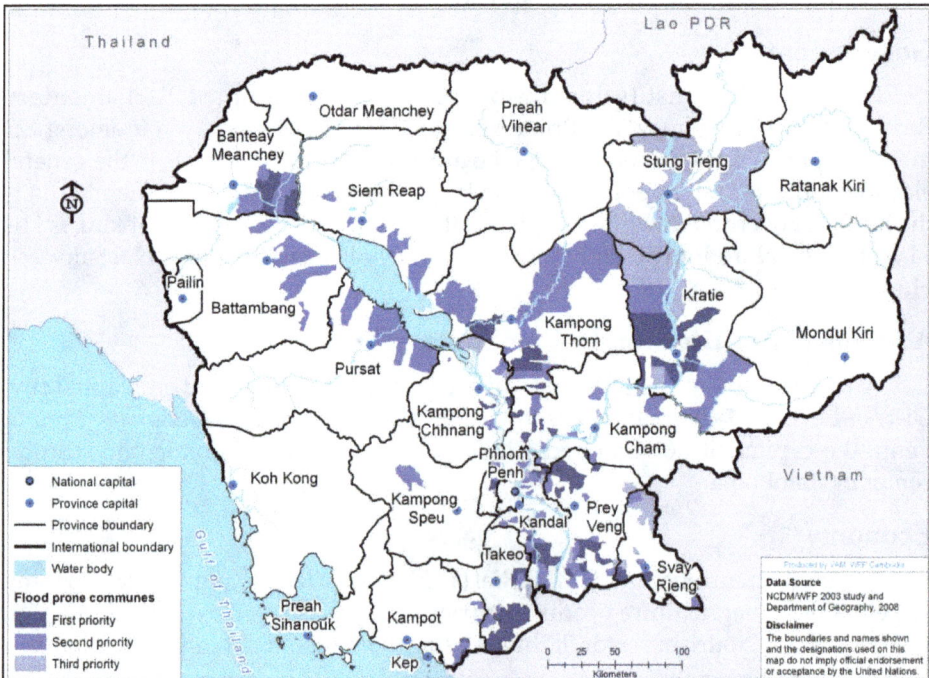

Pursat and Rattanakiri are regularly hit by flash flooding. The second type of flood, the much slower but prolonged flooding, is caused by the overflow of Tonle Sap river and Mekong tributaries, inundating the provinces of Kampong Cham, Kratie, Kandal, Prey Veng, Stung Treng, Svay Rieng and Takeo.

Droughts

Compared to floods, drought is arguably less understood and researched making it difficult to generate national and international response. The drought condition – primarily a result of erratic rainfall – is exacerbated by limited coverage of irrigation facilities (the current coverage is around 20 per cent).

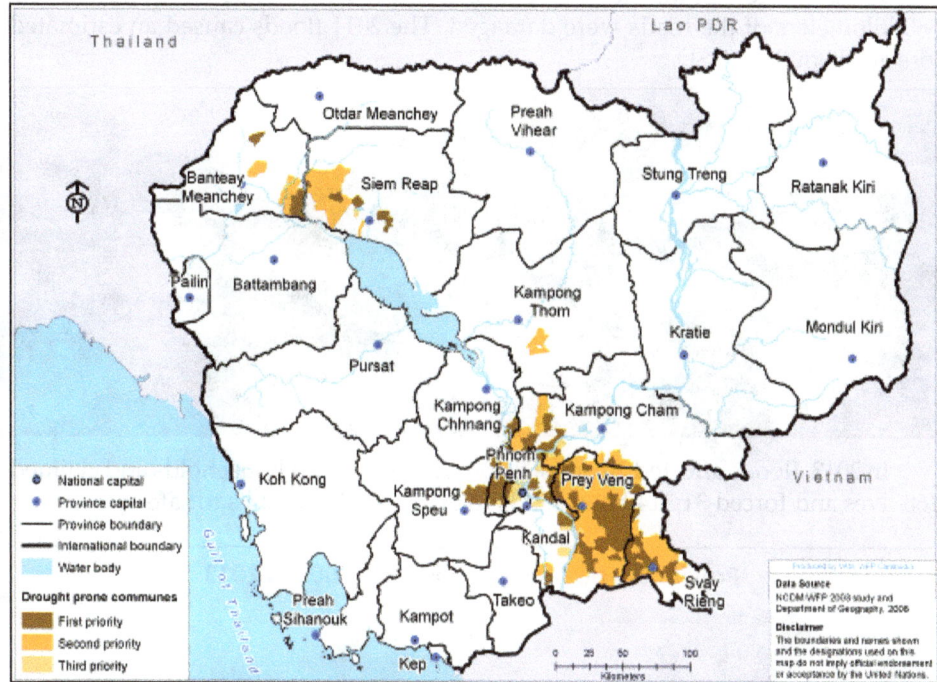

Drought in Cambodia is characterized by loss of water sources caused by the early end or delays in expected seasonal rainfall. Drought severely affects farming productivity especially among rice growing communities who rely solely on rain or river-fed irrigation. Low agricultural yield due to extended drought has increased indebtedness of families and contributed to widespread food shortages.

Typhoons

Tropical cyclones are the most costly meteorological disasters affecting East Asia and the Pacific with, on average, 27 tropical cyclones affecting some parts of the region each year (Chan 2008, quoted in WB 2013). Some typhoons and tropical depressions that reach Indochina do not weaken over the land and produce torrential rainfall and extensive flooding in Cambodia. Typhoon becomes most damaging when it hits during the flooding season (September-October) as it causes heavy precipitation events.

Recent Major Disasters

Flood in 2011 and 2013

In 2011, floods affected 350,000 households (over 1.5 million people) and 52,000 households were evacuated. 18 out of 24 provinces in Cambodia were affected; 4 provinces along Mekong River and Tonle Sap were worst hit. 250 people lives were dead and 23 people sustained injuries in the floods in 2011. 431,000 hectares of transplanted rice fields were affected and 267,000 hectares of rice fields were damaged.

925 kilometers of the national, provincial and urban roads were affected and 360 kilometers of the roads were damaged. The 2011 floods caused an estimated loss at 630 million USD.

In 2013, floods affected 20 out of 24 provinces, 377,354 households and claimed 168 lives and forced 31,314 households to evacuate themselves to safe areas.

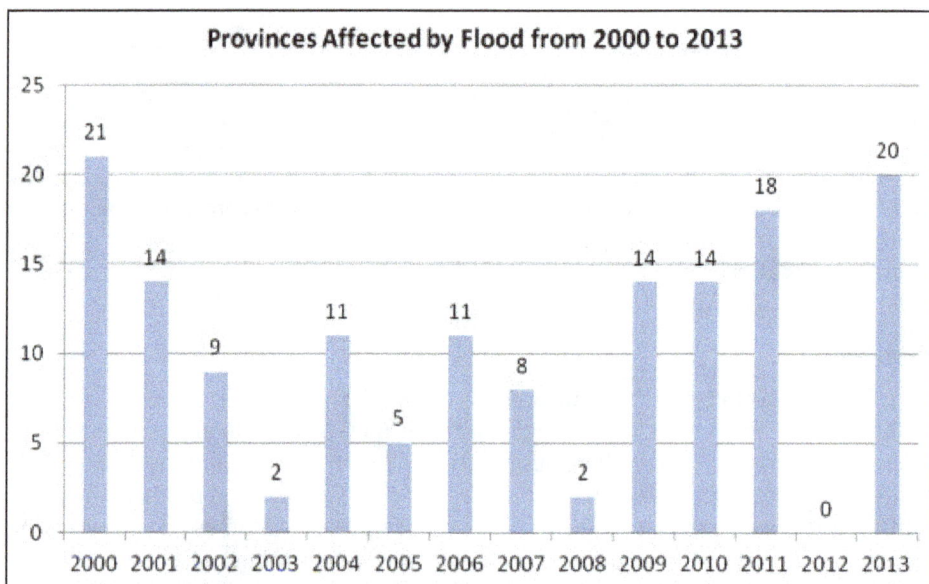

Provinces Affected by Flood from 2000 to 2013

Year	Provinces
2000	21
2001	14
2002	9
2003	2
2004	11
2005	5
2006	11
2007	8
2008	2
2009	14
2010	14
2011	18
2012	0
2013	20

Source: NCDM.

Compared to floods in 2011, floods in 2013 appear to have been less extensive in scale, although in some provinces the impact – including number of evacuated families, damaged crops, damaged infrastructure – was more significant due to a combination of factors such as: unexpected gravity of the floods, both in extent and intensity, longer time for waters to recede, repeated floods and flash floods, limited preparedness undertaken in advance and limited early warning.

Droughts from 2009-2012

In 2009, 13 provinces out of 24 provinces were affected by severe droughts. 57,965 hectares of rice crops were affected and 2,621 hectares were destroyed.

In 2010, 12 provinces out of 24 provinces were affected by severe droughts. 14,103 hectares of transplanted rice were affected by droughts; 3,429 hectares of transplanted rice seedlings and 5,415 hectares of subsidiary crops were damaged.

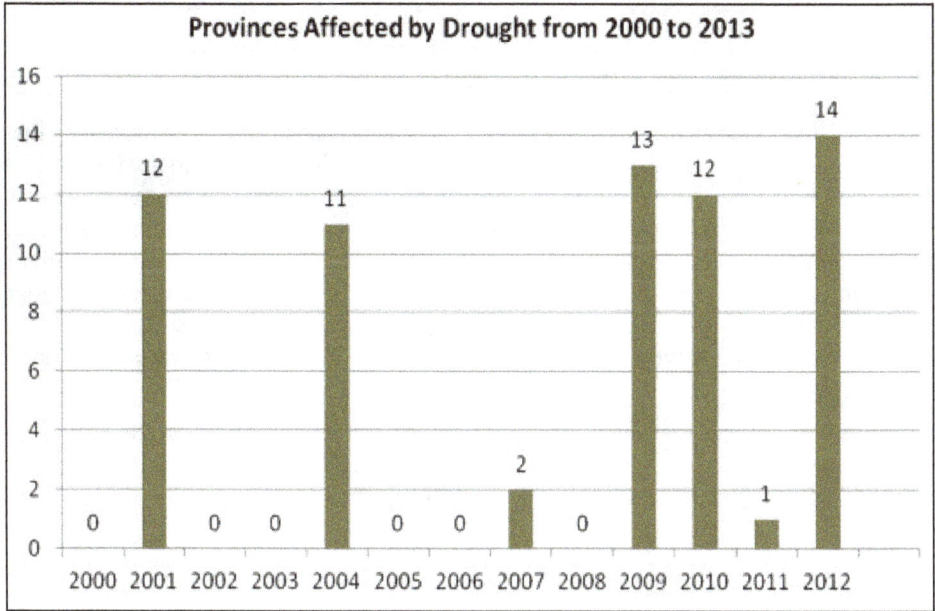

Provinces Affected by Drought from 2000 to 2013

Year	2000	2001	2002	2003	2004	2005	2006	2007	2008	2009	2010	2011	2012
Provinces	0	12	0	0	11	0	0	2	0	13	12	1	14

Source: NCDM.

In 2011, drought affected 3804 hectares of rice fields and destroyed 53 hectares. In 2012, drought hit 11 out of 24 provinces, affected 14,190 hectare of rice fields and destroyed 3151 hectares.

Typhoon Ketsana in 2009

On 29 September 2009, Cambodia was hit by Typhoon Ketsana. 14 out of 25 provinces were hit by the typhoon, and it affected 180,000 households, killed 43 people and injured 67 people.

Lightning Strikes

Lightning strikes claim human lives and livestock and destroyed house and facilities mainly in the rural areas. In 2011, Lightning strike killed and injured 165 and 149 and 101 people and injured 72 people in 2012.

Evolution of Disaster Management

The Royal Government of Cambodia considers disaster management as a key component of its social and economic planning. Floods and droughts have caused serious damage and loss to Cambodia, and endanger the Royal Government's efforts to enhance the economy and well-being of Cambodian society. Cambodia's resources are very limited and these have to be shared across a wide range of development programs such as roads and bridges, and relief for affected communities. It is clear that natural calamities have worsened poverty in Cambodia and thus effective disaster management would be an important contribution to poverty reduction.

Following the adoption of Hyogo Framework of Action, Cambodia developed the National Action Plan and Strategy on Disaster Risk Reduction 2008-2013 to deepen its efforts to reduce disaster risks. The following diagram presents the country's disaster management policy development:

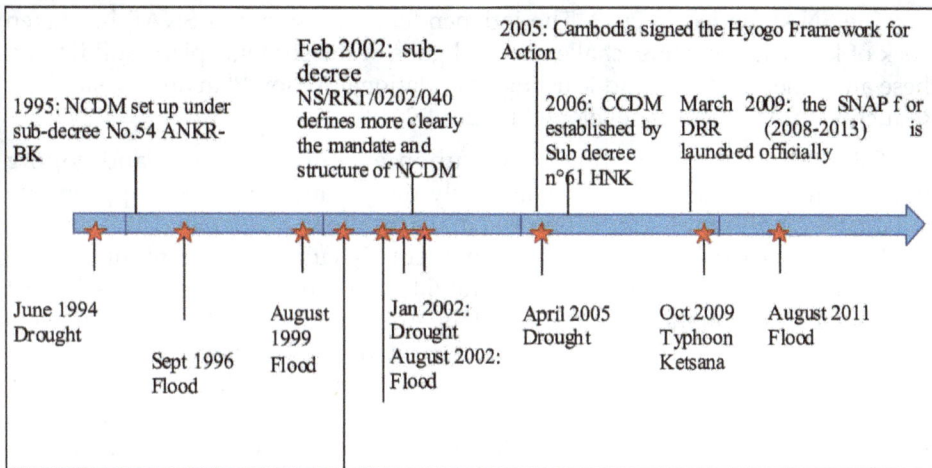

1995: NCDM set up under sub-decree No.54 ANKR-BK

Feb 2002: sub-decree NS/RKT/0202/040 defines more clearly the mandate and structure of NCDM

2005: Cambodia signed the Hyogo Framework for Action

2006: CCDM established by Sub decree n°61 HNK

March 2009: the SNAP for DRR (2008-2013) is launched officially

June 1994 Drought

Sept 1996 Flood

August 1999 Flood

Jan 2002: Drought
August 2002: Flood

April 2005 Drought

Oct 2009 Typhoon Ketsana

August 2011 Flood

Evolution of Disaster Management Policy Development (1994-2011).

Disaster Managment Plan, Policy and Strategy

Legal System and Framework

NCDM Mechanism structure has been established from the national level down to the commune and village levels in conformity with Sub-decree No. 30 ANKR. BK, dated April 09, 2002 on the Organization and Functioning of the National and Sub-National Committees for Disaster Management; Sub-decree No. 61 ANKR. BK, dated June 29, 2006 on the establishment of the Commune Committee for Disaster Management (CCDM); Direction No. 315 NCDM, dated July 21, 2010 on the establishment of the Village Disaster Management Team (VDMT) for the implementation of CBDRM.

Law on Disaster Management was drafted and has been submitted to the National Assembly for enactment by 2015 + National Strategy Development Plan (NSDP) 2009-2013; Strategic National Action Plan for Disaster Risk Reduction (SNAP) 2008-2013 and National Action Plan for Disaster Risk Reduction 2014-2018 (drafted and adopted by 2014)

National Strategic Development Plan (NSDP) Update 2009-2013 and the Strategic National Action Plan on Disaster Risk Reduction (SNAP) 2008-2013 work as the overarching frameworks, and provide strategic direction to disaster risk management for the country. The Royal Government of Cambodia (RGC) has invested considerably to reduce disaster risks through its regular development programmes at national and sub- national levels.

Following the establishment of National Committee on Disaster Management (NCDM) in 2005 and adoption of SNAP in 2008, the RGC has emphasized systematic and proactive efforts in DRR. The country has achieved remarkable knowledge and skills to live with disaster risk. Cambodia endorsed the Hyogo Framework for Action (HFA) in 2005 that provides a systematic and strategic approach to reduction of vulnerability and risk to disasters.

The SNAP ended in 2013. The independent assessment of SNAP has taken stock of key achievements, challenges and gaps of the strategic plan. Building on these and other evidence and learning, this National Action Plan on Disaster Risk Reduction (NAP-DRR) is developed for 2014-2018.

CBDRM is a strategy that builds upon existing capacities and coping mechanisms of communities to collectively design and implement appropriate and doable long-term risk reduction and disaster preparedness plans. The strategy involves the participation of local actors, particularly vulnerable communities, who actively work to identify causes of vulnerability and actions to mitigate the impact of vulnerability from these natural disasters. Additionally, the strategy empowers communities towards long-term capacity to adapt. With recurrent drought and flooding and threats from other natural disasters in Cambodia, CBDRM is seen as the way forward in minimizing enormous loss of life, property and livelihood. In Cambodia, the government considers CBDRM as an integral part of its rural development program to alleviate poverty.

Structure of Disaster Management

National Committee for Disaster Management (NCDM) is established for overall coordination for disaster management. Cambodia has set up necessary

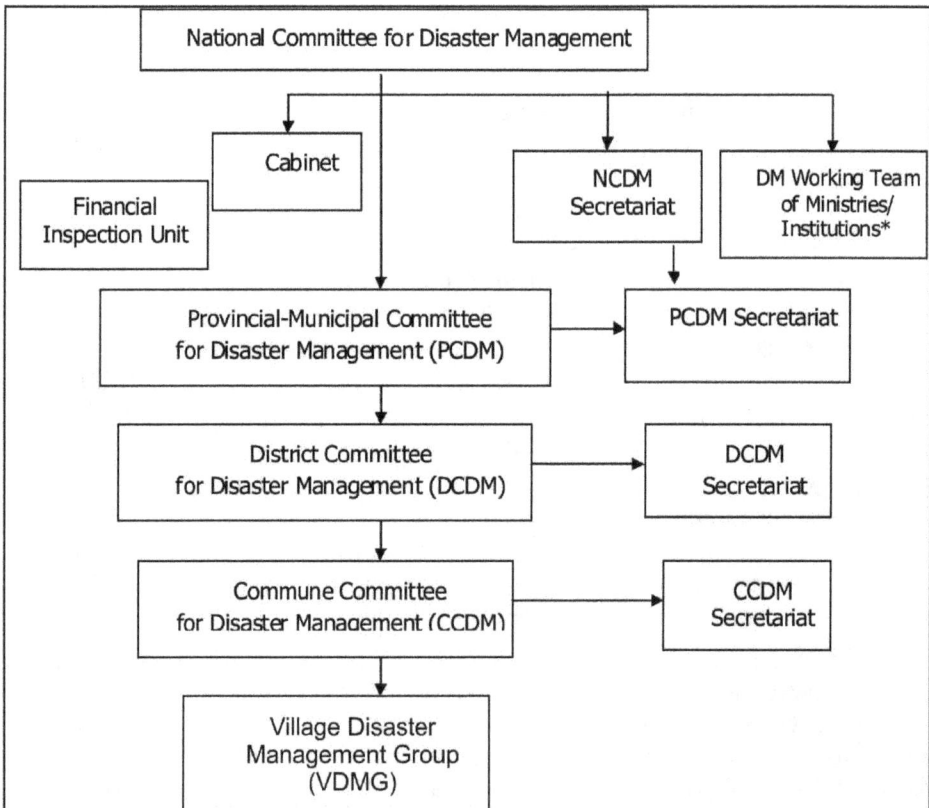

legal, policy and institutional foundations for disaster management. Headed by the Prime Minister, NCDM was set up in 1995 that consists of 22 members from different Ministries, Cambodian Armed Forces, and Civil Aviation Authority as well as representatives of Cambodian Red Cross.

The NCDM Secretariat, which is the locus of disaster management for the country, was set up to lead and coordinate disaster management affairs and to provide support to NCDM. As part of the decentralization process, disaster management institutions such as Provincial Committee for Disaster Management (PCDM), District Committee for Disaster Management (DCDM) and Commune Committee for Disaster Management (CCDM) have been set up to lead disaster management at their respective levels. Village Disaster Management Group (VDMG) is also in place as the lowest level body for disaster management.

DM Working Team of Ministries/Institutions

Each ministry/institution has established the Disaster Working Group of the Ministry/Institution in order to boast the spirit of self-reliance in participating and solving disaster. It is responsible for coordinating all activities involving Disaster Preparedness, Response and Rehabilitation.

Roles and Responsibilities of NCDM General Secretariat

NCDM General Secretariat has the following roles and responsibilities:

- ☆ To ensure the continuity and functioning of the National Committee for Disaster Management administration

- ☆ To conduct research into the flood, drought, storm, wildfire, epidemics prone areas and other hazards and prepare Preparedness and Emergency Response plans

- ☆ To instruct the provincial/municipal, district, commune, committee for Disaster Management and relief communities about work and technical skill that are the basis for collection of disaster data for damage and need assessment and prepare rehabilitation and reconstruction programmes of damaged infrastructure in co- ordination with institutions, UN agencies, IOs, and NGOs concerned

- ☆ To formulate a technical skill training programme for officials who serve Disaster Management functions in provinces, municipalities, district, precinct, and relief communities within the framework of training in and out of the country

- ☆ To coordinate work with Ministries/Institutions concerned, local authorities, UN agencies, IOs, and NGOs in order to evacuate vulnerable people to haven and to provide them with security, public education, Emergency Response and other programmes

- ☆ To give opinion of the documents related to Disaster Management and the letters of consent.

- ☆ To sum the report up and submit it to the National Committee for Disaster Management.

NCDM Secretariat General Organizational Chart

Secretariat General

Department of Administration and Finance	Department of Information and Relations	Department of Emergency Response and Rehabilitation	Department of Preparedness and Training	Department of Search and Rescue
Administration Bureau	Information and Prediction Bureau	Emergency Bureau	Training Bureau	Searching Coordination Bureau
Accounting Bureau	International Relations Bureau	Rehabilitation Bureau	Planning Bureau	Rescue Coordination Bureau
Logistic Bureau	Newsletter and Publishing Bureau	Operation Bureau	Program Bureau	

National Committee for Disaster Management

Departments under NCDM General Secretariat

The General Secretariat of the National Committee for Disaster Management has 5 **departments**. The organizational chart below illustrates the departments under the General Secretariat and line bureaus of each department.

Progress of the Implementation of Hyogo Framework of Action (HFA)

The progress to date of implementation of each of HFA priorities of action is listed below:

1. **Priority for Action 1: Ensure that disaster risk reduction is a national and local priority with a strong institutional basis for implementation**

   ☆ Finalization and improvement the draft of disaster management law to be submitted for approval.

   ☆ Establishing and strengthening disaster management mechanisms by determining the organization and functioning of the national, sub-national, and local levels to match with the actual situation.

   ☆ Developing the policy guideline, legal instrument and legal framework to support the disaster risk reduction activities.

   ☆ Mainstreaming the disaster risk reduction into policy guidelines and development plans at all levels.

2. **Priority for Action 2: Identify, assess and monitor disaster risks and enhance early warning**

   ☆ Conducting the risk assessment at national, sub-national, and local levels.

   ☆ Developing the vulnerability and hazard maps.

   ☆ Developing the disaster data management system.

   ☆ Recording, analyzing, and disseminating the disaster losses information.

   ☆ Setting up the early warning system (EWS) on hazards for Sangkat-commune, particularly communities exposed to hazards.

   ☆ Designating the focal points for the Emergency Operation Centre (EOC) at national and sub-national levels.

   ☆ Developing the capacity of technological research and analyse, forecasting natural hazards map and other hazards which vulnerable to the disaster impacts.

   ☆ Improving the existing data for further assessment, monitoring and early warning in conformity with the regional and international levels.

   ☆ Strengthening capacity of recording, analyzing, summation, disseminating, and exchanging information, statistics, and general methodology for hazard assessment and monitoring.

3. **Priority for Action 3: Use knowledge, innovation and education to build a culture of safety and resilience at all levels**

   ☆ Collecting, compiling, and disseminating the knowledge and information on hazards, vulnerabilities, and capacities to the people in order for building the culture of prevention and disaster resilience.

   ☆ Providing simple, understandable and protective information on disaster risk to the people who exposed to hazards.

   ☆ Developing the Sangkat-commune disaster risk reduction plans.

   ☆ Strengthening the cooperation and promoting the partnership among the relevant stakeholders including the professional in socio-economy for disaster risk reduction.

   ☆ Promoting the utilization, implementation and accessing new information, communication, space technology and other services, including the interpretation of satellite maps.

   ☆ Developing a guide manual on disaster risk reduction for utilization at local community, sub-national and national levels.

   ☆ Preparing the international standard terminologies on disaster risk available in national language for utilization in specialized institutions, training materials, and public awareness programmes.

   ☆ Integrating the disaster risk reduction concept and disaster preventive programme into the curriculums of school and higher education institution.

4. **Priority for Action 4: Reduce the underlying risk factors**

   ☆ Mainstreaming the disaster risk reduction related to climate change into the strategy on disaster risk reduction and climate change adaptation.

   ☆ Mainstreaming the disaster risk reduction plan into the health, education, agriculture, forestry and fishery, and rural development sectors.

   ☆ Promoting hazard resilience of communities in the droughts, floods, storms, and other hazards prone areas which caused the people's livelihoods declined.

   ☆ Strengthening the disaster recovery plan including the socio-psychology training programme in reducing the adverse impact on victim mentality, particularly women and children in post-disaster.

5. **Priority for Action 5: Strengthen disaster preparedness for effective response at all levels**

   ☆ Developing the preparedness plan for emergency response and updating the contingency plan to be effective at all levels.

   ☆ Forming up the emergency response coordinating teams.

   ☆ Forming up the search and emergency rescue teams.

   ☆ Forming up the disaster assessment coordinating teams.

   ☆ Developing a proper coordinating procedure inconformity with ASEAN Coordinating Centre for Humanitarian Assistance on Disaster Management (AHA Centre) and implementing the ASEAN Agreement on Disaster Management and Emergency Response (AADMER) effectively.

   ☆ Promoting the disaster preparedness simulation exercise, including the real exercise and victim evacuation in order to ensure the rapid and response and timely receiving the important relief to the needs of the affected localities.

   ☆ Reserving fund for the emergency response.

   ☆ Building the safe areas by equipping with bathrooms, latrines, and shelters for human beings and animals.

Recent Major Projects on Disaster Risk Reduction

Project Implementation and Bird Flu Prevention (Avian and Human Influenza Control and Preparedness Emergency Project -AHICPEP)

This project was funded by the World Bank and implemented by three institutions with different resposibities, but with the same purpose and goal of supporting implementation of the national plan on Avian and Human Influenza. The three involved institutions are comprised of the Ministry of Agriculture, Forestry and Fisheries, responsible for animal health, the Ministry of Health, responsible for human health and NCDM, responsible for inter-ministerial cooperation for preparedness and prevention of epidemics and pandemics and coordination of implementation of the common plan. The project came into effect on 06 August 2008 and was concluded on 31 December 2011.

Implementation of Post-Ketsana Reconstruction Project

The Ministry of Rural Development and NCDM implemented the post ketsana, immediate restoration and reconstruction project (World Bank loan). The goal of the project was the restoration of the physical infrastructure damaged by Ketsana storm in 2009, including rural roads, bridges, cultverts, hand pump wells and lavatories in the provinces of Siem Reap, Banteay Meanchey, Battambang, Kampong Cham, Kampong Thom and Kampong Chhnang for the period of five years, starting from 2011 to 2014, implemented by the Ministry of Rural Development. The project also dealt with capacity in disaster management of the Royal Government of Cambodia, regarding effective prevention, emergency response and reconstruction after disasters, implemented by NCDM.

Disaster Risk Reduction and Climate Change Adaptation Cooperation for 2013-2015 between NCDM and Caritas Cambodia

The project was aimed at promoting disaster risk reduction and strengthening the preparedness, prevention, emergency response and rehabilitation as well as in building capacity of the committee's officials at sub-national level.

Strengthening National and Sub-national Capacity to Implement Disaster Management towards Increased Community Resilience

Funded by the European Commission Directorate General for Humanitarian aid and Civil Protection (DG ECHO), the National Committee for Disaster Management (NCDM) in collaboration with Plan International Cambodia (Plan) and World Vision Cambodia (WVC) are currently implementing the CONSORTIUM PROJECT "Strengthening National and Sub-national Capacity to Implement Disaster Management towards Increased Community Resilience"(Project Duration: 1st of June 2012 to the 30th of November 2013). The goal of the project is to increase resilience of vulnerable people including children and women living in the most disaster-affected areas in Cambodia. The project's main objective is to contribute to an "Enhanced coordination and increased collaboration among NCDM and partners resulting from the development of partnership guidance (ToR, Action Plan, *etc.*) and disaster management related documents".

Disaster Risk Reduction (DRR) Forum

Coordinated and chaired by NCDM and attended by various stakeholders working in the field of disaster management, DRR forum is held quarterly, aiming to enhance disaster management in Cambodia through improved information sharing, coordination of initiatives and joint action to promote the highest possible standards and disaster risk reduction practices to reduced suffering and losses of the affected population from disaster.

Conclusion and Recommendation

Cambodia is one of the five countries located along the Mekong River and dominated by monsoon; rainy and dry season. The country experiences hydro-meteorological hazards such as floods, drought, storms (typhoon), fire incidents and epidemics. During the monsoon season, Cambodia experiences flash floods usually after heavy rainfall. The second type of flood, the much slower but prolonged flooding, is caused by the overflow of Tonle Sap river and Mekong tributaries. According to National Committee for Disaster Management (NCDM), in 2011, floods affected 350,000 households (over 1.5 million people) and 52,000 households were evacuated. 250 people lives were dead and 925 kilometers of the national, provincial and urban roads were affected and 360 kilometers of the roads were damaged. The 2011 floods caused an estimated loss at 630 million USD. Drought severely affects farming productivity especially among rice growing communities who rely solely on rain or river-fed irrigation. In 2012, drought hit 11 out of 24 provinces, affected 14,190 hectare of rice fields and destroyed 3151 hectares.

The challenges and implications of disaster management in case of Cambodia is facing such as lack of appreciation on disaster risk management institutions; lack of understanding of the importance of database/information based decision making of the decision maker; lack of appreciation and commitment to database and disaster information management promotion and use; inadequate resources, manpower, professions and skills; understanding of different level stakeholders (politician, technical planner and workers) database; very limitation of resource allocation for Disaster Management Information System, (lack of human and financing resource), and systematic procedures and cooperation among NCDM, all line agencies and NGOs in implementation of DRM are not compatible. To urgently deal with this problems, the Royal Government of Cambodia has set up future scenarios and concrete steps taken for DRR and considers disaster management as a key component of its social and economic planning. Cambodia developed the National Action Plan and Strategy on Disaster Risk Reduction to deepen its efforts to reduce disaster risks. Cambodia has set up necessary legal, policy and institutional foundations for disaster management. Headed by the Prime Minister, NCDM was set up in 1995 that consists of 22 members from different Ministries, Cambodian Armed Forces, and Civil Aviation Authority as well as representatives of Cambodian Red Cross. The NCDM Secretariat, which is the locus of disaster management for the country, was set up to lead and coordinate disaster management affairs and to provide support to NCDM. Additionally, the country still need donor Commitment to the development of national capacity on disaster information management and database; supports from national and regional institutions on the process of capacity building; TA, hardware and software; support country level building strong Government's institution for disaster information management and databases; advocate to decision maker to use database and reliable disaster information for decision making; capacity Building to National and Sub-National level. (Training of Trainer, National to Sub-National level), looking for any projects: Pilot projects on Geo-referrence Information System for DRM, Drought Monitoring, and Climate Change Pilot Project and necessary equipments.

As part of the decentralization process, disaster management institutions have been set up at Sub-National Level to lead disaster management at their respective levels. Therefore, Royal Government's effort is to enhance the economy and well-being of Cambodian society. It is clear that natural calamities have worsened poverty in Cambodia and thus effective disaster management would be an important contribution to poverty reduction.

References

1. HE Samdech Hun Sen, 2004. *Address at the Sixth International Meeting of Asian Disaster Reduction Center and the Third Meeting of the Secretariat of International Strategy for Disaster Reduction*, Phnom Penh, 02 February 2004.

2. Abhas K. Jha, *et al.*, 2013. *Strong, Safe, and Resilient A Strategic Policy Guide for Disaster Risk Management in East Asia and the Pacific. World Bank. 2013.*

3. Humanitarian Response Forum (HRF) Final Report - No. 07, Dec 2013. *Cambodia: Floods.*

4. NCDM, 2011. Summary Annual Report on Disaster Events in Cambodia from 2000-2010.

5. NCDM *et al.*, 2008. *Strategic National Plan for Disaster Risk Reduction, 2008-2013.*

6. NCDM, 2013. *National Action Plan for Disaster Risk Reduction (NAP-DRR) 2014-2018 (draft).*

7. HE. Mr. Ponn Narith, 2013. *HFA Implementation: Progress Report for ACDR2014.*

8. HE. Mr. MA Norith, 2012. *Presentation on Cambodia's Disaster Impacts, National Committee for Disaster Management, Cambodia.*

9. HE. Mr. Ross Sovann *et al.*, 2010. *Cambodia Presentation on Climate Risk Information Workshop*, Bangkok.

10. ADPC *et al.*, 2008. *Monitoring and Reporting Progress on Community Risk Management in Cambodia.*

Chapter 10

A Synopsis of Disaster Management in South Africa: An Intergovernmental Approach to Disaster Prevention, Mitigation and Response

Dumisani Emmanuel Mthembu

Department of Science and Technology,
Private Bag x 894, Pretoria
E-mail: dumisani.mthembu@dst.gov.za

ABSTRACT

This paper looks at the policy and regulatory framework for disaster management in South Africa and how the structure of the South African government has shaped the disaster management approach; by decentralising and institutionalising disaster management in all spheres of government. South Africa is vulnerable to disasters, including those caused by climate change and its related extreme weather events. Therefore, South Africa is not immune to the impacts of climate change. It is for this reason that disaster management is increasingly becoming one of the great challenges of the 21st century due to socio-economic losses emanating from natural or human-made disasters. As such, preventing and responding to disaster requires a systematic approach to the management of its impacts. In South Africa, impacts from natural hazards matter a lot because of the increasing levels of disaster risk it faces. This is aggravated by its geographical location, coupled with extensive coastline. As a result, disaster management in South Africa developed dramatically since 1994 and provided a paradigm shift from centralised civil protection and reactive approach, to integrated and coordinated disaster risk management with special emphasis on prevention and mitigation. This paper highlights the key features of the policy and legislative framework for disaster management. It further provides a critical analysis of the policy and legislative framework for disaster management. It concludes that the legislation provides adequate measures and mechanisms for disaster risk reduction, prevention and recovery. However, a lot more still needs to be done to improve the implementation and build the requisite capacity at all spheres of government.

Keywords: Disaster management, South Africa, Integrated, Coordinated, Prevention, Mitigation, National, Provincial, Local.

Introduction

Climate change is one of the greatest environmental challenges and the most pressing phenomenon of our time. It is commonly accepted that in South Africa, climate change related hazards and the resulting disasters are on the increase particularly in urban areas. It also remains one of the greatest threats to sustainable development, and has the potential to undo or undermine the positive advances made in meeting South Africa's development goals; and will (if unmitigated) certainly thwart the attainment of the post-2015 development agenda. Research has shown that while the impacts of climate change will affect all regions, however, the Intergovernmental Panel on Climate Change (IPCC) report of 2007 concluded that Africa is one of the most vulnerable continents to climate variability and change because of multiple stresses and low adaptive capacity.

According to Vermaak and van Niekerk (2004), South Africa is not prone to spectacular, destructive and media-attracting disasters such as volcanic eruptions. Disasters have mainly been dominated by localised incidents of fire, informal settlement fires, seasonal flooding in vulnerable communities, droughts, oil spills and mining incidents. However, Faling, Tempelhoff and van Niekerk (2012) have observed that South Africa faces increasing occurrences of floods, tornadoes, and hailstorms, storms, heavy rain and winds, veld fires, snow and drought. They further suggest that urban areas in South Africa are particularly vulnerable to climate change related disasters where structural poverty, substandard infrastructure and housing, high population density, economic assets and industrial activities are concentrated. The impact of climate change on human settlements ranges from insignificant to catastrophic. Municipalities in developing countries, including in South Africa, experience major setbacks in hard-won economic and social development following such events. Governments in general have insufficient capacity to predict, monitor, mitigate and manage hazards and disasters especially those that are induced by climate change (Faling, Tempelhoff and van Niekerk, 2012).

Drought is a major disaster in South Africa in terms of total economic loss and number of people affected, whilst floods top the chart in terms of the number of mortalities (Ngaka, 2012). Disaster management in South Africa has traditionally been the domain of individual government Departments and voluntary organisations (Vogel, 1998). It has also been largely reactive and centralised. Vogel (1998) has also noted that coordinating between Departments and disaster management structures was poor. Furthermore, the focus previously was largely on relief and response with no attention to prevention and mitigation. As a result, Vermaak and van Niekerk (2004), have argued that the field of disaster management developed dramatically since 1994, resulting in paradigm shift from centralised civil protection and reactive approach to integrated and coordinated disaster risk management with special emphasis on prevention and mitigation. These changes have also resulted in a comprehensive legislative framework that moved away from traditional disaster response thinking, supported by a decentralised approach by incorporating disaster risk reduction in the hierarchical structure in all spheres of government.

An Overview of Past Climate Related Disasters in South Africa

As indicated earlier, South Africa is vulnerable to a number of extreme weather events ranging from floods, drought, fires and large storms. These events have serious socio-economic implications for the country, particularly, cost related to recovery and economic growth. The impacts are wide-ranging and affect a number of sectors. The impacts also include primary or direct effects, such as loss of infrastructure and death and secondary or indirect effects which include health issues and loss of livelihoods (Easterling, Meehl, Parmesan, Changnon, Karl and Mearns, 2000). In South Africa, the poor are the most vulnerable because of their limited access to livelihood opportunities, infrastructure, information, technology and assets. This is regrettably aggravated by lack of adequate planning, implementation and insurance cover for disaster losses (Botha, Van Niekerk, Wentik, Tshona, Maartens, Forbes, Annandale, Coetzee and Raju. 2011; Botha and Van Niekerk 2013).

Like with any other disasters, climate related disasters cause extensive damage to the economy and the society resulting in increase in recovery costs. For example, South Africa's second national communication to the United Nations Framework Convention on Climate Change revealed that between 2000 and 2009, the estimated total cost of weather related events totalled approximately R9.2 billion. Similarly, the annual report of the National Disaster Management Centre (NDMC) of 2011 revealed that from 2006 to 2011, the provincial and district municipalities shouldered the greatest financial liability from weather related events totalling approximately R1 596 billion incurred for rehabilitation and recovery in the 2010/2011 financial year. These astonishing figures could even be higher due to inadequate data of the national extent of the impacts. There is a need to make sure that data collection is improved and consistently recorded uniformly especially on costs of disasters.

Until 2014, the 1991/1992 drought was ranked as the worst natural disaster in South Africa. In 2015, South Africa experienced the epic drought in 30 years, and is still reeling from the devastating effects of this drought. According to Mniki (2009), the 1991/1992 drought resulted in approximately 70 per cent of crops failing and South Africa was forced to import maize. The spill over effects of the drought resulted in 50 000 jobs being lost in the agricultural sector and further 20 000 from related sectors. This was due to increased debt and farm closures resulting from the drought. According to Pretorius and Small, (1992), the loss of the Gross Domestic Product (GDP) was approximately 1.8 per cent which represented $500 million.

Another common climate related disaster in South Africa is floods. Previous experience has shown that while floods have caused the highest economic cost of damages, droughts tend to affect a larger proportion of the country's population. For instance, between 1900 and 2014, it is estimated that 17 million people (34 per cent of the population) have been affected by drought while floods have affected approximately 570 000 people. In 1987, the KwaZulu Natal province experienced unprecedented flooding which was the second most severe flooding which destroyed kilometres of road infrastructure, washed away 14 bridges and closed all entry points to Durban, which is one of the major cities in KwaZulu Natal.

According to Grobler (2003), at least 68 000 people were left without homes, and 388 people died. In 2003 and 2005 respectively, the Province of the Western Cape experienced floods that was attributed to the cut off lows, causing R260 million in damages (Mukheibir and Ziervogel, 2007). In November 2016, serious flooding (flash floods) took place in Johannesburg (Gauteng Province) which resulted in the destruction of property and infrastructure. Few deaths were also recorded. As a result, a local disaster was declared. It is also important to note that statistics on flooding is not comprehensive as flooding which regularly occur in informal settlements and high-risk low lying areas is not consistently recorded. These are flood prone areas that experiences localised flooding, leaving people stranded and vulnerable to disease outbreaks.

Wildfires are another common occurrence particularly in Mpumalanga Province due to long dry season. In 2002, at least 24 000 of pasture was destroyed by fire and four people were recorded dead. According to Forsyth, Kruger and Maitre. (2010), the costs amounted to R32 million. Between 2007 and 2008, significant losses of revenue were experienced both in KwaZulu Natal and Mpumalanga provinces amounting to $430 million. During these fires, 61 700 hectares of plantation forest were lost, equating to 2.9 per cent and 9.5 per cent of the total area of plantations in these two provinces. Most of the timber was unfortunately unsalvageable resulting in huge economic losses.

The coastal areas in South Africa are susceptible to storm surges. These often results in damage to infrastructure including coastal properties. According to Smith, Guastella, Mather, Bundy, and Haigh. (2013), the damage to coastal properties and infrastructure in 2007 and 2008 was estimated to be R 1 billion. In 2007, about 350 kilometres of coastal soil erosion was as a result of storm surges in KwaZulu Natal (Breetze, Parak, Celliers, Mather and Colenbrander, 2008). It is clear from this discussion that South Africa has a fair share of climate induced disasters with considerable socio-economic costs. This makes it even more imperative to have a proactive system in place to anticipate and deal with disasters in general.

Materials and Methods

The review in this paper was based on document analysis. According to Bowen (2009), document analysis is a systematic procedure for reviewing or evaluating documents, both printed and electronic materials. This methodology requires that data be examined and interpreted in order to produce meaning, gain understanding with a view to develop empirical knowledge. Relevant South African government policy documents relating to disaster management were purposefully searched and selected, particularly the Disaster Management Act (DMA) and the National Disaster Management Framework (NDMF). Further information was obtained from qualitative analysis of literature, the websites of the Disaster Management Institute of Southern Africa, and the National Disaster Management Centre of South Africa as a means of triangulation in order to seek convergence and corroboration through the use of different data sources. The analysis included an iterative process of skimming, reading and interpretation; combined with thematic analysis. Thematic analysis is a form of pattern recognition within data, with emerging themes becoming the categories for analysis (Bowen, 2009).

Results and Discussion

Cooperative Governance and Institutional Arrangements

After the June 1994 floods on the Cape Flats (Western Cape province), the Cabinet resolved to assess South Africa's ability to deal with risks and disaster management (Vermaak and Van Niekerk, 2004). This resulted in the review of disaster management structures and policies in South Africa, including the legislative process. Van Niekerk (2014) has concurred that the legislative and policy-making process was initiated following severe flooding in the Western Cape. The policy review was also necessitated by a need for a paradigm shift in traditional disaster response thinking, to disaster risk reduction, prevention and mitigation (Van Niekerk, 2014). Furthermore, the policy review was in response to the fragmented nature of disaster management which resided in different government Departments and entities prior to 1994. As such, South Africa had to develop a policy and regulatory framework for disaster management that would provide for a coherent and coordinated approach to disaster management nationally.

Vermaak and Van Niekerk (2004) provides an extensive analysis of the process that was followed in the development of the new legal regime and the institutional arrangements. For instance, in 1995, the Cabinet recommended that a formal structure for disaster management be established. The then Provincial and Local Government (now Department of Cooperative Governance and Traditional Affairs) was designated as an interim focal point for disaster management. The National Disaster Management Committee was formed in 1996 to coordinate and manage disaster management policy formulation.

In 1997, the Inter-Ministerial Committee for disaster management was formed consisting of ten national Departments. Vermaak and Van Niekerk also notes that the task team emerged from the Inter-Ministerial Committee in 1998, which oversaw the process of developing and tabling the Green Paper (draft policy) on disaster management in all spheres of government. In order to deal with South Africa's immediate needs, an Interim Disaster Management Centre was established in 1997, which was later replaced by the National Disaster Management Centre (NDMC) that was established under the DMA.

The institutional arrangements for disaster management in South Africa are informed by the governance framework prescribed by the Constitution (1996). According to the Constitution, disaster management is a functional area of concurrent national and provincial legislative competence. The Constitution is the supreme law of the country and provides a broad framework on how the government must be structured to enable it to function properly. The government is constituted as national, provincial and local spheres of government which are distinctive, interdependent and interrelated. As a result, over and above the national sphere of government, there are also nine provincial governments and 278 municipal councils. The configuration of government has relevance on how disaster management is undertaken in South Africa. The Constitution also enjoins all spheres of government to cooperate with one another, while remaining distinctive.

In 2002, the South African government enacted the Disaster Management Act (27 of 2002). According to Vermaak and van Niekerk (2004), the DMA gives guidance to the legal establishment of the NDMC, the duties and powers of the national, provincial and local spheres of government, and funding for post-disaster recovery and rehabilitation. The purpose of DMA is to provide for an integrated and coordinated disaster management policy that focuses on preventing or reducing the risk of disasters, mitigating the severity of disasters, emergency preparedness, rapid and effective response to disasters and post disaster recovery; provide for the establishment of national, provincial and municipal management centres as well as disaster management volunteers. In 2005, the South African NDMF was promulgated, in response to the DMA.

The institutional arrangement and governance system of disaster management in South Africa is crucial; given the structure of government and the inherent separation of powers between the different spheres of government. These institutional arrangements are important to ensure effective and efficient disaster management preparedness as well as response given the different responsibilities and duties by different role players. The DMA establishes a number of institutions such as the (i) the Intergovernmental Committee on Disaster Management (ICDM) which is formed by relevant Cabinet Ministers appointed by the President, nominated Members of Executive Council from provincial governments and municipal councils; (ii) the National Disaster Management Advisory Forum (NDMAF) constituted by the Minister; as well as provincial and local forums (iii) the NDMC, and (iv) Local Disaster Management Centres. All of these institutions play an important role by exercising their powers and duties in preventing, mitigating and responding to disasters.

The ICDM is chaired by the Minister. The ICDM reports to Cabinet on the coordination of disaster management among all spheres of government and must ensure that the principles of cooperative government as prescribed in the Constitution are applied. Furthermore, the ICDM provides advice and make recommendations to Cabinet on issues related to disaster management and on the establishment of a national framework for disaster management. Such a framework must ensure an integrated and uniform approach to disaster management by all spheres of government, statutory bodies, non-governmental organisations involved in disaster management, the private sector, communities and individuals. The schematic institutional framework is depicted in Figure 10.1.

The Minister is also required to establish a NDMAF. The Forum is a body in which national, provincial and local government and other disaster management role players consult one another and coordinate their actions on matters related to disaster management. The Forum makes its recommendations concerning the

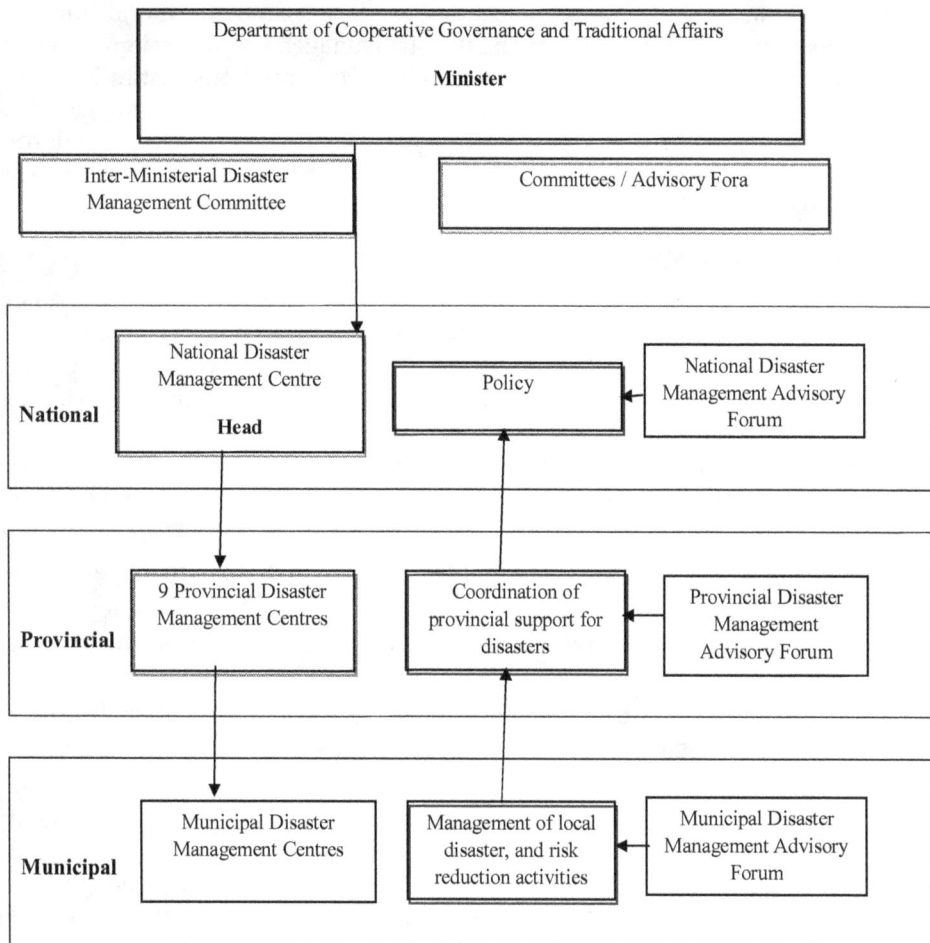

Figure 10.1: Institutional Arrangements and Responsibilities of Disaster Management in South Africa (adapted from the NDMF).

NDMF to the ICDM. The Forum may also provide advice to any organ of state, statutory body, non-governmental organisation, community or private sector on any matter relating to disaster management.

The Forum consist of a variety of stakeholders and is chaired by the Head of the NDMC. The Forum consist of senior representative of each national Department whose Minister is a member of the ICDM; a senior representative of each provincial Department whose Member of Executive Council (MEC) is a member of the ICDM designated by that MEC as well as municipal officials selected by the South African Local Government Association. These representatives represent all three spheres of government. In addition, the Forum consist of representatives from disaster management role players including: organised business, the chamber of mines, organised labour, the insurance industry, organised agriculture, traditional leaders, religious and welfare organisations, medical, paramedical and hospital organisations, organisations representing disaster management professions in South Africa, relevant non-governmental and international organisations, institutions that can provide scientific and technological advice or support to disaster management, institutions of higher education and statutory bodies regulating safety standards in particular industries. This makes the Forum representative, multi-disciplinary and multi-sectoral.

National Disaster Management Framework

According to Vermaak and van Niekerk (2004), a further structure established by the DMA is the NDMF. The Minister must prescribe the NDMF. The NDMF must among others be informed by the recommendations of the ICDM as well as public comments. Its purpose is to provide a coherent, transparent and inclusive policy on disaster management for the country. It must also reflect a proportionate emphasis on disasters of different kinds, severity and magnitude that occur or may occur in Southern Africa and place emphasis on measures that reduce the vulnerability of disaster-prone areas, communities and households. It must further:

☆ Guide the development and implementation of disaster management in South Africa;

☆ Establish prevention and mitigation as the core principles of disaster management;

☆ Facilitate South Africa's cooperation in regional and international disaster management and the establishment of joint standards of practice;

☆ Give effect to the application of cooperative governance issues concerning disasters and disaster management among the spheres of government by determining the relationship between the sphere of government exercising primary responsibility of the coordination and management of disaster and spheres of government performing supportive roles; as well as allocating specific responsibilities to the different spheres;

☆ Guide the development and implementation of disaster management within national, provincial and municipal organs of state on a cross-functional and multi-disciplinary basis and allocate responsibilities accordingly;

☆ Facilitate the participation of other disaster management role players;

☆ Facilitate disaster management capacity building, training and education;

☆ promote disaster management research;

☆ Guide the development of the comprehensive information management system; and

☆ Provide the framework within which organs of state may fund disaster management with specific emphasis on preventing or reducing the risk of disasters, including grants to contribute to post-disaster recovery and rehabilitation and payment of victims of disasters and their dependants.

The provincial and local spheres of government must similarly establish and implement a framework for disaster management aimed at ensuring an integrated and uniform approach to disaster management by all in the respective spheres. Such framework must be consistent with the provisions of the Act as well as the NDMF.

National Disaster Management Centre

The DMA makes provision for the establishment of the NDMC as an institution within government reporting to the Minister. Its objective is to promote an integrated and coordinated system of disaster management, with a special emphasis on prevention and mitigation, by national, provincial and municipal organs of state, statutory bodies and other role-players involved in disaster management. The Minister appoints the Head of the NDMC whose job is to ensure that the NDMC exercise its powers and performs its duties. All decisions making powers rests with the Head of the NDMC except those that have been delegated or assigned to provincial or local levels in terms of section 14 of the DMA. Having said that, the Head may confirm, vary or revoke any decision taken as a result of the assignment.

The NDMC is responsible for taking all measures necessary to achieve the objectives of the DMA. This includes:

- ☆ Specialising on issues concerning disasters and disaster management;
- ☆ Monitoring whether organs of state and statutory bodies comply with the Act, NDMF and monitors progress with post-disaster recovery and rehabilitation;
- ☆ Act as a repository of, and conduit for information concerning disasters, impending disasters and disaster management;
- ☆ May act as an advisory and consultative body;
- ☆ Make recommendations regarding funding of disaster management and initiate and facilitate efforts to make such funding available;
- ☆ Make recommendations on draft legislation affecting the Act, or any other disaster management issue; on the alignment of national, provincial and municipal legislation with this Act or the NDMF; in the event of national disaster, on whether a national state of disaster should be declared;
- ☆ Promote research in disaster management as well as capacity building, training and education.

Similarly, each province and each metropolitan or district municipality must establish a Provincial Disaster Management Centre (PDMC) or a Municipal Disaster Management Centre (MDMC), as the case may be in its administration for its area. The PDMC forms part of and functions within a Department designated by the Premier in the provincial administration. Both the provincial and municipal centres must be headed by the Head duly appointed to exercise the powers and performance of duties conferred to the centre.

The DMA also make provision for the development of disaster management plans and strategies. All spheres of government must also develop the Disaster Management Plans and strategies which must be updated on a regular basis. The

NDMC must develop guidelines for their preparation, and regular review and updating by organs of state and other institutional role-players involved in disaster management. It must also assist in aligning these plans and strategies and the implementation of these plans and strategies by all respective organs of state and role-players. Every year, the NDMC must submit an annual report to the Minister on its activities and performance of its functions and duties. A copy of such a report must also be made available to provincial and municipal disaster management centres. The Minister must then submit the report to Parliament within 30 days from date of receipt of the report.

Given the multiplicity of stakeholders involved in disaster management, the National Centre must develop and maintain a directory of institutional role-players that are or should be involved in disaster management in Southern Africa, establish effective communication links and further act as a repository of, and conduit for information concerning disasters and disaster management and monitor the implementation and compliance with the Act. This is done through the disaster management information system. The NDMC is also empowered to gather relevant information from spheres of government and other role players as deemed appropriate.

Classification and Recording of Disasters

In the event of the disastrous event occurring or threatening to occur at a national level, the NDMC must determine whether the event meet the criterion of being classified as a disaster in terms of the Act, assess the magnitude and severity or potential thereof, classify the disaster as a local, provincial or national disaster; as well as record the prescribed particulars concerning the disaster in the prescribed register. In assessing the magnitude and severity of the disaster, the NDMC must consider any information and recommendations concerning the disaster received from a provincial or municipal disaster management centre. It may also enlist the assistance of an independent assessor to assess the disaster on site.

These determinations can also be made by the relevant structures at a provincial or local level; hence the need for cooperative governance. Where this is the case, the provincial or municipal disaster management centre as the case may be, must also inform the NDMC of the disaster and its initial assessment of the magnitude and severity or potential magnitude or severity of the disaster; and further make such recommendations regarding the classification of the disaster as may be appropriate. The NDMC may reclassify a disaster classified as a local, provincial or national disaster after consultation with relevant provincial or municipal disaster management centres if the magnitude and severity of the disaster is greater or lesser than the initial assessment.

The disaster is classified as a local disaster if it affects a single metropolitan, district or local municipality, and if that municipality alone is able to deal with it effectively; or with the assistance of local municipalities in the area of the district municipality. A disaster is classified as a provincial disaster if it affects more than one metropolitan or district municipality in the same province; or a single metropolitan or district municipality in the province and that metropolitan or district municipality is unable to deal with it effectively. Importantly, the province concerned must be able to deal with the disaster effectively.

National disasters affect more than one province or a single province which is unable to deal with it effectively. The Minister may by notice in the Gazette declare a national state of disaster if the existing legislation and contingency arrangement do not adequately provide for the national executive to deal effectively with the disaster or other special circumstances warrant the declaration of a national state of disaster. The Minister may also make regulations or issue directions or authorise the issue of directions concerning:

☆ The release of available resources of the national government, including stores, equipment, vehicles and facilities;

☆ The release of personnel of a national organ of state for the rendering of the emergency services;

☆ The implementation of all or any provisions of the national disaster management plan that are applicable under the circumstances;

☆ The evacuation to temporary shelters of all or part of the population from the disaster-stricken or threatened area if such action is necessary for preservation of life;

☆ The regulation of the movement of person and goods, from or within the disaster-stricken or threatened area;

☆ The control and occupancy of premises in the disaster-stricken or threatened area;

☆ The provision, control or use of temporary emergency accommodation;

☆ The suspension or the limiting of the sale, dispensing or transportation of alcoholic beverages in the disaster-stricken or threatened area;

☆ The maintenance or installation of temporary lines of communication to, from or within the disaster area;

☆ The dissemination of information required for dealing with disaster;

☆ Emergency procurement procedures;

☆ Facilitation of response and post-disaster recovery and rehabilitation; and

☆ Other steps that may be necessary to prevent an escalation of the disaster or to alleviate, contain and minimise the effects of a disaster including steps to facilitate international assistance.

Until a disaster has been classified accordingly, it is deemed a local disaster. The classification of a disaster designates primary responsibility to a particular sphere of government for the coordination and management of the disaster, but an organ of state in another sphere may assist the sphere that has primary responsibility to deal with the disaster and its consequences. The national state of disaster lapses three months after it has been declared. It may also be extended by the Minister by notice in Gazette.

Funding of Post Disaster Recovery and Rehabilitation

This classification of disasters is crucial to facilitate funding, including for post-disaster recovery and rehabilitation. In South Africa, public funds are managed in accordance with the Public Finance Management Act (PFMA), which also provides for the use of funds for emergency situations. When a disaster occurs, national, provincial and local organs of state may financially contribute to response efforts and post-disaster recovery and rehabilitation. The cost of repairing or replacing public sector infrastructure is borne by the organ of state responsible for its maintenance.

Upon declaration and classification of a disaster, emergency funds may be released for this purpose and national government may also make contributions to alleviate effects of local and provincial disasters. Any financial assistance provided must be in line with the NDMF and any applicable post-disaster recovery and rehabilitation policy. The funding must take into account:

☆ Whether any prevention and mitigation measures were taken;

☆ Whether the disaster could have been avoided or minimised had prevention and mitigation measures been taken;

☆ Whether it is reasonable to expect that prevention and mitigation measures should have been taken in the circumstances;

☆ Whether the damage caused by the disaster is covered by adequate insurance; and

☆ The extent of financial assistance available from community, public or other non-governmental support programmes; and the magnitude and severity of the disaster, the financial capacity of victims of the disaster and their accessibility to commercial markets.

Perspective on Effectiveness of Implementation

Various scholars in disaster risk management have commended the progressiveness and robustness of the DMA and the NDMF (decentralisation, its focus on disaster risk management and emphasis on sectoral and community engagement). It has been argued that from an international perspective, the contents of the DMA and NDMF are sound in terms of contemporary disaster risk reduction, however, there have been constraints and challenges in implementation. According to van Niekerk (2014), the implementation is hindered by the lack of a strong institutional basis. Even though the DMA is explicit about various institutional arrangements needed for effective disaster risk management, Botha *et al.* (2011 and van Riet and Diedericks (2010) found that these key structures are not always in place. Where they are in place, they do not tend to function adequately or productively (van Niekerk, 2014); hence the need to bolster their capacity, training and education.

Similar challenges have been observed in sectoral Departments. Van Niekerk (2014) argues that not all relevant national Departments have implemented the required disaster risk management activities. Neither have they identified focal points for disaster risk management as required by the DMA. The failure to address disaster risk management at the local municipal level hinders the implementation of the DMA and NDMF. According to Vermaak and van Niekerk (2004), the Integrated Development Plan (IDP) which is a strategic planning and developmental instrument for municipalities in South Africa in terms of the Municipal Systems Act (32 of 2000) explicitly requires the municipalities to reflect on applicable disaster management plans.

However, Schipper and Pelling (2006), has noted that there is a glaring lack of the integration of disaster risk management in development planning at all levels, particularly municipalities which is worrying; taking into account the functions of the municipalities in achieving the developmental objectives. Previous scrutiny of IDPs by van Retief (2009a, 2009b) found very little evidence of such plans containing any reference to disaster risk management, let alone the existence of integrated disaster risk management plans.

van Niekerk (2014) has also noted that, the disaster risk management function at the local government level remains inadequately funded, even though the Municipal Systems Act requires that the assignment of functions to municipalities must be supported by additional financial support. Hence funding for disaster management is severely limited at the municipal level or reprioritised due to competing demands.

Some disaster management practitioners have questioned the placement of disaster management in the Ministry rather than at the Presidency (the highest political office). However, others argued that disaster management necessitates placement of the national centre within an environment that will allow for its effective functioning and the implementation of its mandate and the Presidency may not necessarily be suitable for that; even though it would provide the requisite political support and elevate the importance of disaster management to the highest echelons of power and decision making. Furthermore, all institutional arrangements are replicated at the lower levels with a view to facilitate implementation. This however may create dual centres of responsibility, thus ultimately undermine the authority of the national centre. The local centres may also be hesitant to take decisions, hoping that the national centre will intervene.

Regarding funding, (start-up costs and ongoing disaster risk reduction activities, as well as disaster response and recovery); the mechanisms are not always uniformly applied or implemented by all spheres of government (van Niekerk, 2011). The Disaster Management Institute of South Africa has recommended that Municipal Finance Management Act (Act 53 of 2003) must be amended. In particular, section 29 (2)(b) which states that unforeseeable and unavoidable expenditure during an emergency or disaster may be authorised but that this may not exceed a percentage of the budget. This, they have argued that it restricts the amount of funds available to respond to an emergency or disaster.

The process and procedures for classifying and declaring a disaster has also been criticised. According to van Niekerk (2014), the process and procedures are unclear and tend to be cumbersome. The emboldened role of the NDMC may defeat the very purpose of decentralising disaster management and may ultimately disempower the municipalities from declaring a state of disaster. There is also a potential for duplication of declarations, which occurs once other legislation is invoked such as the Fund Raising Act (107 of 1978).

Conclusions

The weak institutional capacity particularly at local level is of concern. Faling (2008) has concurred that the disaster risk reduction sentiments are not reflected to the same degree of gravity in spatial development planning-related policy and legislation. It is also of concern that planning for climate change and reducing disaster risk are among the most underestimated issues on the agenda particularly at local government level. Even though local governments are torn between attending pressing socioeconomic development priorities and introducing environmental concerns into planning processes, climate change impact considerations and disaster management should be integral to local government planning and other planning processes. Faling (2008) has also correctly observed that where the day-to-day needs of people are scarcely being met, issues of sustainable development are given only momentary attention and are difficult to reconcile with more immediate priorities. Faling *et al.* (2012) has further argued that officials in local government do not appear to understand climate change science and the implications of climate change at local level, hence the relegation of disaster management on the periphery. It must be recognised that given resource and capacity constraints, relevant officials may also be faced with more immediate environmental issues. There are also challenges with insufficient capacity and financial resource, to adapt and incorporate climate change considerations into political and administrative decision making.

While these are real challenges, failure to confront them is problematic and somewhat short-sighted, given the benefits of disaster risk reduction and mitigation. Having said that it must equally be recognised that the South African government has been able to deal with many disasters which otherwise would have been disastrous. While a lot could still be learnt to enhance and improve operational efficiencies, it would appear that the South African model provides for an efficient, effective and yet proactive and pragmatic approach to disaster management. There is also an urgent need to build capacity particularly at local levels and to ensure that disaster management is mainstreamed in planning processes; and importantly, ensure that adequate resources are set aside for disaster management interventions.

Acknowledgements

I am very grateful to the Department of Science and Technology (South Africa), and my colleagues for nominating me to participate in the International workshop on mitigation of disasters due to severe climate events. I also wish to extend my gratitude and to acknowledge the organisers of the workshop: the Centre for Science and Technology of the Non-aligned and other Developing Countries (NAM S&T Centre) and the National Science and Technology Commission (NASTEC) for accepting my nomination, for the support and all logistical arrangements before and during the workshop; including sponsoring my participation and hospitality.

Abbreviations

DMA: Disaster Management Act

ICDM: Intergovernmental Committee on Disaster Management

IDP: Integrated Development Plan

GDP: Gross Domestic Product

MEC: Member of Executive Council

MDMC: Municipal Disaster Management Centre

NDMC: National Disaster Management Centre

NDMF: National Disaster Management Framework

NDMAF: National Disaster Management Advisory Forum

PDMC: Provincial Disaster Management Centre

PFMA: Public Finance Management Act

References

1. Boko, M., I. Niang, A. Nyong, C. Vogel, A. Githeko, M. Medany, B. Osman-Elasha, R. Tabo and P. Yanda, 2007: Africa. *Climate Change 2007: Impacts, Adaptation and Vulnerability. Contribution of Working Group II to the Fourth Assessment Report of the Intergovernmental Panel on Climate Change*, M.L. Parry, O.F. Canziani, J.P. Palutikof, P.J. van der Linden and C.E. Hanson, Eds., Cambridge University Press, Cambridge UK, 433-467.

2. Botha D. *et al.*, 2011. Disaster risk management status assessment at Municipalities in South Africa. Pretoria. SALGA.

3. Bowen, G. A. 2009. Qualitative Analysis as a Qualitative Research method. *Qualitative Research journal, Vol. 9 (2), 27-40.*RMIT Publishing.

4. Faling, W, Tempelhoff, J. W. N and van Niekerk,D. 2012. Rhetoric or action: Are South African Municipalities planning for climate change? *Development Southern Africa*, 42 (29): 241-257.

5. Faling, W. 2008. Vulnerability to disaster impacts: One of the most underestimated issues in urban development. Planning Africa Conference, 14-16 April, Johannesburg.

6. Mattingly, S. 2002. Policy, Legal and Institutional Arrangements. Regional Workshop on Best Practices in Disaster Mitigation. Asian Disaster Preparedness Centre, Bangkok. 25–27.

7. Ngaka, M. K., 2012. Drought preparedness, impact and response: A case of the eastern cape and Free State provinces in South Africa. *Jamba: Journal of Disaster Risk Studies* 4(1).

8. Pelling, M. and A. Holloway. 2006. Legislation for Mainstreaming Disaster Risk Reduction. Teddington, United Kingdom: Tearfund.

9. South Africa. 1978. Fund-Raising Act 107 of 1978. Pretoria: Government Printer.

10. South Africa. 1996. Constitution of the Republic of South Africa. Act 108 of 1996. Pretoria: Government Printer.

11. South Africa. 2000. Municipal Systems Act 32 of 2000. Pretoria: Government Printer.

12. South Africa. 2002. Disaster Management Act, 2002 (Act 57 of 2002). Pretoria: Government Printer.

13. South Africa. 2005. National Disaster Management Policy Framework. Pretoria: Government Printer.

14. South Africa. 1999. Public Finance Management Act, 1999.(Act 1 of 1999). Pretoria.

15. Schipper, L. and Pelling, M. 2006. Disaster risk, climate change and international development: Scope for, and challenges to intergration. Disasters. 30 (1): 19-38.

16. van Niekerk, D. 2014. A critical analysis of the South African Disaster Management Act and Policy Framework. *Disasters*, 38(4): 858–877.

17. van Niekerk, D. 2005. *A Comprehensive Framework for Multi-sphere Disaster Risk Reduction in South Africa*. Ph.D. thesis. Potchefstroom, South Africa: North-West University.

18. van Niekerk, D. 2006. 'Disaster Risk Management in South Africa: The Function and the Activity—Towards an Integrated Approach'. *politeia*. 25(2), pp. 95–115.

19. van Niekerk, D. 2007. Disaster Risk Reduction, Disaster Risk Management and Disaster Management: Academic Rhetoric or Practical Reality?' *Disaster Management: Southern Africa*. 4(1), pp. 6–9.

20. van Niekerk, D. 2008. From Disaster Relief to Disaster Risk Reduction: A Consideration of the Evolving International Relief Mechanism'. *Journal for Transdisciplinary Research in Southern Africa*. 4(2): 355–75.

21. van Niekerk, D. *et al.*, 2011 *Alternative Financing Model for Disaster Risk Reduction in South Africa*. Pretoria: Financial and Fiscal Commission.

22. van Riet, G. 2009a. Disaster Risk Assessment in South Africa: Some Current challenges. *South African Review of Sociology*. 40(2): 194–208.

23. van Riet, G. 2009b Qualitative and Quantitative Community Level Approaches: Disaster Risk Assessment in the Fezile Dabi District Municipality'. *Africanus: Journal of Development Studies*. 39(2): 31–43.

24. van Riet, G. and Diedericks, M. 2010. The pacement of disaster management centres in District, metropolitan municipality and provincial government structures. *Administratio Publica*. 18 (4): 155-73.

25. Vermaak, J. and Niekerk, D. 2004. Development debate and practice: Disater risk reduction in South Africa. *Development Southern Africa* Vol. 21, No. 3, 2004.

26. Vogel, C. 1998. Disaster management in South Africa. *South African Journal of Science*, 94 (3): 98-106.

Chapter 11

Possible Early Warning for Landslide in Sri Lanka using "Antecedent Daily Rainfall Index": A Case Study of Meeriyabedda Landslide on 29th October 2014

W.N.S. Rupasinghe[1] and K.H.M.S. Premalal[2]

[1]*Meteorologist,*
Department of Meteorology, Colombo, Sri Lanka
E-mail: mcksenali@yahoo.co.in
[2]*Director,*
Department of Meteorology,
Colombo, Sri Lanka
E-mail: spremalal@yahoo.com

ABSTRACT

Landslide is one of the hazardous events which cause lives and property damages. Many factors, such as soil type, slope of the terrain, precipitation and manmade activities incorporate for landslides. Frequent of occurring landslide is increasing in Sri Lanka due to the erratic rainfall pattern with climate change. National Building Research Organization (NBRO) is the national focal point for identifying locations, early warnings and mitigation.

Meeriyabedda landslide in the Koslanda estate in Badulla district occurred in 29th October, 2014 at about 7.45am. This severe catastrophe event caused 12 deaths and 23 disappearances, completely wiping out all the state line houses. Victims of the Meeriyabedda landslide disaster are estate workers residing in line houses.

Study was carried out to find a suitable method for identify possibility of early warning using antecedent daily rainfall index. Antecedent daily rainfall indexwas applied for 5 landslides which occurred in Badulla district and the constant in the equation for antecedent daily rainfall index was identified as 0.9.

The results give good indication for landslide with the existing rainy condition even few days early that the event occurred. Therefore use of this equation for many landslide cases and fine tune the results will be benefitted for future early warnings.

Keywords: Landslide, Rainfall intensity, Antecedent daily rainfall.

Introduction

Landslide is one of the hazardous events which cause lives and property damages. Structure of the soil and the intense and continuous rainfall are the main reasons for landslides but it can be triggered due to manmade activities over the landslide vulnerable areas. However, landslide prone areas were identified by National Building Research Organization (NBRO) in Sri Lanka by studying various factors.

Meeriyabedda landslide in the Koslanda estate in Badulla district occurred in 29th October, 2014 at about 7.45am (Somarathne, 2014). This severe catastrophe event caused 12 deaths and 23 disappearances, completely wiping out all the state line houses. Victims of the Meeriyabedda landslide disaster are estate workers residing in line houses (Somarathne, 2014).

The rainfall measured at Poonagala rainfall measuring station which is the closest station to Merriyabedda, was 8.0 mm on 29th October, but according to observations, rainy conditions were prevailed during whole month except few days. However continuous rainy spells were started on 14th October and 784mm rainfall were received until 29th October 2014. This amount is much higher compare to the climatological mean.

Month of October is belong to the second intermonsoon period and heavy thunderstorms and high intense rainfall can be expected almost over the all island. Climatologically, higher rainfall receives during the period from October to December in Badulla District (Figure 11.1). As the Meeriyabedda area is prone to the landslide, period of October to December are vulnerable months for landslides. The other reason is the high intense rain accompanied thunderstorms.

Many research studies have been conducted to correlate landslide with the precipitation pattern. Early work by Caine (1980) established rainfall duration-intensity thresholds(Glade, 2000). Since then numerous other research groups have worked on landslide-triggering rainfall thresholds (Caine, 1980; Cannon and Ellen, 1985, Ellen and Wieczorek, 1988; Harp *et al.*, 1997; Keefer *et al.*, 1987; Wilson *et al.*, 1993; Wilson and Wieczorek, 1995; Wieczorek, 1987; Church and Miles, 1987).

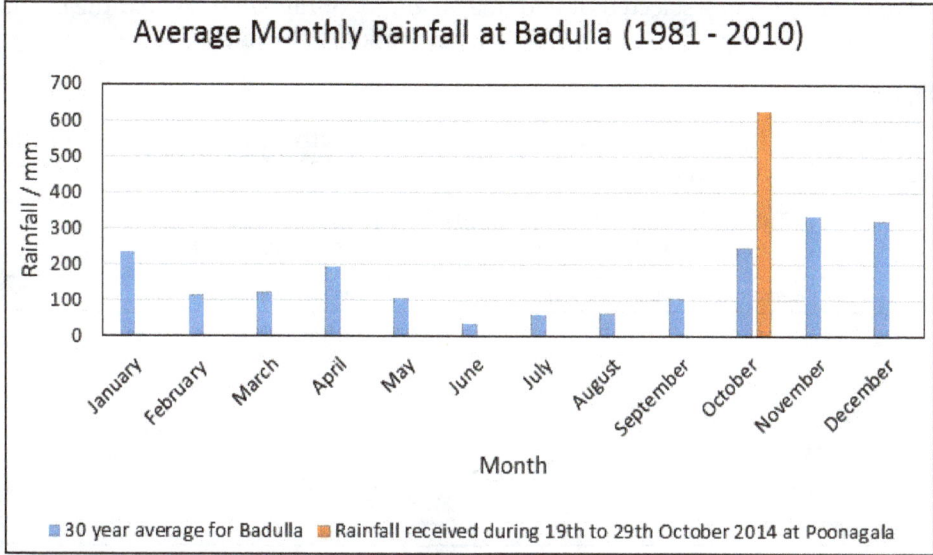

Figure 11.1: 30 Year Average Rainfall for Badulla District and the Rainfall Received at Poonagala Rainfall Station from 19th to 29th October 2014.

Rainfall duration, rainfall intensity, cumulative event rainfall, and antecedent rainfall are the most commonly investigated variables (Giannecchini et,al. 2012). In addition, one day heavy precipitation event also cause landslides. In particular, landslide initiation is frequently related to rainfall intensity and duration(Caine, 1980; Aleotti, 2004; Giannecchini, 2006; Guzzetti etal., 2007, 2008; Cannon *et al.*, 2008; Coe *et al.*, 2008; Dahaland Hasegawa, 2008).

The "Antecedent Daily Rainfall Index" (Glade *et al.*, 2000) was used for the analysis for the landslide events in Badulla district. It is a power equation (equation 1) and R_1 to R_n represents the daily rainfall for the past n days prior to the landslide event. Then the R_0 is the antecedent daily rainfall index on zerothday, that is the day of occurring landslide. k is an arbitrary constant (0<k<1) and the suitable value for k can be guessed by applying the equation for historical landslide events at a place where landslide occurred. It is needed to calculate the antecedent daily rainfall indices for the day occurrence of landslides. The value for the k can be identified by construct a line which passes more events in the graph of the antecedent rainfall indices and the rainfall on the zeroth day.

$$R_0 = kR_1 + k^2R_2 + k^3R_3 + k^4R_4 + k^5R_5 + k^6R_6 + \ldots\ldots k^nR_n \tag{1}$$

Crozier and Eyles (1980), following Bruce and Clark (1966) used k=0.84, which comes from Ottawa (United States) stream flow. Akatsu, 2010 identified the value of k as 0.8 by considering historical landslide events in Ratnapura area in Sri Lanka for 6 day antecedent period.

In this study analysis has been done for 6 days and 10 days rainfall prior to the date of landslide and the previous date of landslide in Badulla district, since there were few cases selected due to unavailability of rainfall data near the event locations and in case of high rainfall was not the source of landslide.

Data and Methodology

Historical landslides event on the Badulla district was collected (Table 11.1) from the National Building Research Organization (NBRO).

Table 11.1: Historical Landslide Events in Badulla District
(*Source*: National Building Research Organization)

Place of the Landslide Event	Data and the Year
Agarathenna	1986.01.10
Viharagala	1992.11.16
Passara- Namunukula road	1993.12.18
Welimada	2004.12.18
Galahitiyawa	2006.12.20
Meeriyabedda	2014.10.29

Daily rainfall data were obtained from the Department of Meteorology at the nearest rainfall measuring station to the point where landslide occurred. Since, only 6 cases are considered in the Badulla area, previous day also considered as probable landslide due to following reasons.

1. No rainfall measuring stations are at the same location. Therefore the rainfall may be different at the place of event occur.

2. There may be some slow movement (some indication) even at the previous day to the landslide occur.

Equation is applied for 6 days and 10 days daily rainfall prior to the landslide. Part of the study is to identify a suitable k value. Therefore study was conducted for different k values as 0.2, 0.3, 0.4, 0.5, 0.6, 0.7, 0.8, 0.85, 0.9 and 0.95.

All the landslides shown in the Table 11.1 is occurred in October, November, December period where the highest rainfall received during the year due to 2nd inter - monsoon and North-East monsoon.

Results and Discussion

The scatter plot of daily rainfall with antecedent daily rainfall index is shown in the figure 2 and 3 for landslides in Badulla district for 6 days and 10 days rainfall prior to the event. The arbitrary line was constructed in order to cross much events.

☆ The value of the x axis, where the arbitrary line meets is the antecedent daily rainfall index for past 6 days (for 6 day period) and 10 days (for 10 day period) even without any rainfall for the day of event.

☆ The value of the y axis, where the arbitrary line meets is the daily rainfall threshold even without any rainfall for the last 6 days and 10 days.

(Note : It is needed to be careful, because the value of y axis should be the threshold value for one day rainfall even without rain during past days)

According to the NBRO, the value for one day rainfall for landslide is 150 mm, if it is not an intense rain. Considering this information, line was constructed to cross y axis for the value near to 150. Finally following results were obtained.

More landslide events were not captured by the line for k=0.20 to 0.60. Better indication has been given for k=0.70 – 0.85, but the threshold value for one day rainfall is more than 200mm. It is not follow the NBRO criteria. The value for one day threshold is 200mm for k= 0.90 to 0.95. Even the NBRO criteria for landslide is 150mm per day, it does not show from the analysis. Analysis given it as 200 mm or little above. Antecedent daily rainfall index also vary between 150 and 200 mm (Figure 1(d) to (j)). For better analysis study was extended for 10 day rainfall analysis to calculate antecedent daily rainfall index.

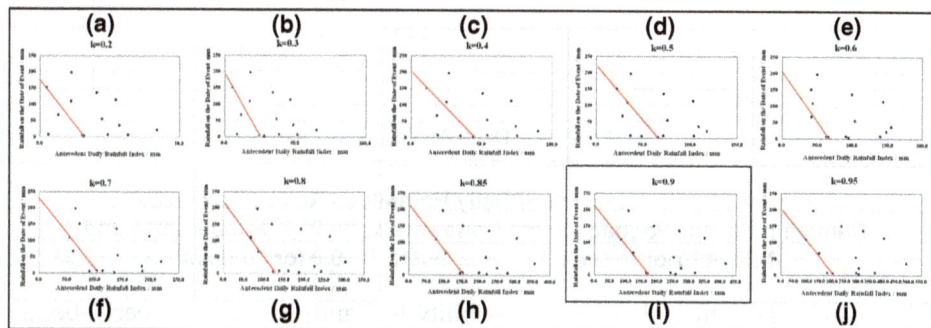

Figure 11.2: Scatter Plot of Daily Rainfall with Antecedent Daily Rainfall Index for different k Values (6 day).

Results are almost similar as the Figure 11.2. When compare both figures, graph of k = 0.90 captures most of the events therefore, antecedent daily rainfall index can be taken as 200 mm as the best threshold value for Badulla district. Similar to 6 day analysis, threshold value of daily rainfall is little above 200 mm. Figure 11.4 shows the scatter plot of daily rainfall with antecedent daily rainfall index for 4 days prior to the landslide and the day of landslide at Meeriyabedda on 29th October 2014 with k=0.90.

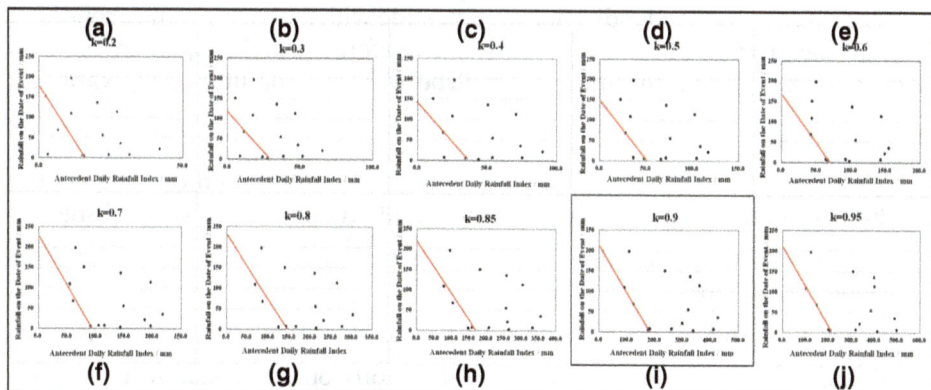

Figure 11.3: Scatter Plot of Daily Rainfall with Antecedent Daily Rainfall Index for different k Values (10 day).

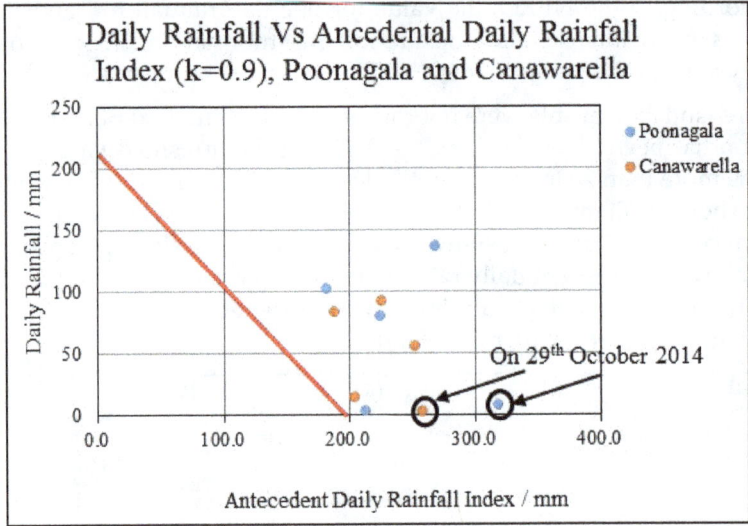

Figure 11.4: Scatter Plot of Daily Rainfall Vs Antecedent Daily Rainfall Index for Poonagala and Canewarella on 29th October 2014 and 4 Days Prior to the Landslide Occur (k=0.9 for 10 days).

It can be seen that, there was possibility for landslide at the Meeriyabedda area even for few days prior to the landslide. Therefore when there's continues rainfall, it is better to consider antecedent rainfall index as the threshold value, 150.0 mm - 200.0 mm.

Conclusion

Analysis has been conducted to obtain the preliminary idea about the possibility of finding a method for early warning. Some hydrological model should be applied for better prediction. Study of rainfall intensity give a good indication for landslides occur due to high intense rainfall. Therefore it is important to establish many automated rain gauge systems to monitor in-situ rainfall.

The results give good indication for landslide with the existing rainy condition even few days early that the event occurred. Therefore use of this equation for many landslide cases and fine tune the results will be benefitted for future early warnings.

References

1. Akatsu, K, 2010. Study and Presentation on 'Identification of Landslide in Ratnapura, Sri Lanka, using Antecedent daily rainfall index' Unpublished.

2. Aleotti, P., 2004. A Warning System for Rainfall-induced Shallo Failuers, Eng. Geol., 73, 247-265.

3. Bruce, J. P., and Clark, R. H., Introduction to Hydrometeorology, 317 p.

4. Caine, N. 1980. The Rainfall Intensity-duration Control of Shallow Landslides and Debris Flows,Geografiska Annaler A 62 (1–2), 23–27.

5. Canon, S.H., and Ellen, S. 1985. Rainfall Conditions for Abundant Debris A6alanches SanFrancisco Bay Region California, California Geology (December 1985), 267–272.

6. Cannon, S. H., Gartner, J. E., Wilson R. C., Bowers J. C., and Laber J. L., 2008. Storm rainfall conditions for floods and debris flows from recently burned areas in southwestern Colorado and southern California, Geomorphology, 96, 250–269.

7. Church, M. and Miles, M.J., 1987. Meteorological antecedents to debris flows in southwestern BritishColombia; Some case studies. In Debris Flows:A6alanches: Process, Recognition, and Mitigation (J. E.Costa, and G. F. Wieczorek, eds.) Reviews in Engineering Geology (The Geological Society ofAmerica, Boulder 1987) pp. 63–80.

8. Coe, J. A., Kinner, D. A., and Godt, J. W. 2008. Initiation conditions for debris flows generated by runoff at Chalk Cliffs, central Colorado, Geomorphology, 96, 270–297.

9. Crozier, M. J., and Eyles, R. J., 1980. Assessing the probability of rapid mass mo6ement. In The NewZealand Institution of Engineers—Proceedings of Technical Groups (ed.). Proc. Third Australia–NewZealand Conference on Geomechanics, Wellington, 2.47–2.51.

10. Dahal, R. K. and Hasegawa, S. 2008. Representative rainfall thresholds for landslides in the Nepal Himalaya, Geomorphology, 100, 429– 443.

11. Ellen, S. D., and Wieczorek, G. F., 1988. Landslides, Floods, and Marine Effects of the Storm of January 3 –5, 1982, in the San Francisco Bay Region, California, U.S. Geological Survey Professional Paper 1434, 1–283.

12. Giannecchini, R. 2006. Relationship between rainfall and shallow landslides in the southern Apuan Alps (Italy). *Nat. Hazards Earth Syst. Sci.*, 6, 357–364.

13. Giannecchini, R., *et al.,* 2012. Critical Rainfall Threshold for Triggering Shallow Landslides in the Serchio River Valley (Tuscany, Italy), Natural Hazards and Earth System Sciences, 12, 829-842.

14. Glade, T., Crozier, M., Smith, P., 2000. Applying Probability Determination to Refine Landslide-triggeringRainfall Thresholds Using an Empirical "Antecedent Daily Rainfall Model" Pure applied Geophysics, 157, 1059–1079.

15. Guzzetti, F., Peruccacci, S., Rossi, M., and Stark, C. P. 2007. Rainfall thresholds for the initiation of landslides in central and southern Europe, *Meteorol. Atmos. Phys.*, 98, 239–267.

16. Guzzetti, F., Peruccacci, S., Rossi, M., and Stark, C. P. 2008. The rainfall intensity-duration control of shallow landslides and debris flows: an update, Landslides, 5, 3–17.

17. Harp, E. L., Chleboard, A. F., Schuster, R. L., Cannon, S. H., Reid, M. E., and Wilson, R. C., 1997. Landslides and Landslide Hazards in Washington State due to February 5 –9, 1996 Storm, U.S. Geological Administrative Report, Denver, 28 pp.

18. Keefer, D. K., Wilson, R. C., Mark, R. K., Brabb, E. E., Brown III, W. M., Ellen, S. D., Harp, E. L., Wieczorek, G. F., Alger, C. S., and Zatkin, R. S., 1987. Real-time Landslide Warningduring Heavy Rainfall, Science 238 (13 November 1987), 921–925.

19. Somarathne, M. 2014. Challenges to Overcome : An Overview of Recent Ladslides. *News Letter, Geological Society of Sri Lanka*, Vol. 31, 07-10 pp.

20. Wieczorek, G. F., Effect of rainfall intensity and duration on debris flows in central Santa CruzMountains, California, In Debris Flows:A6alanches: Process, Recognition, and Mitigation (J. E. Costaand G. F. Wieczorek, eds.) (The Geological Society of America, Boulder 1987) pp. 93–104.

21. Wilson, R. C., Mark, R. K., and Barbato, G., 1993. Operation of a Real-time Warning System forDebris Flows in the San Francisco Bay Area, California, Hydraulic Engineering, ASCE, Hydraulics Division, ASCE, San Francisco, California, 1908–1913.

22. Wilson, R. C., and Wieczorek, G. F. 1995. Rainfall Thresholds for the Initiation of Debris Flows at La Honda, California. *Environm. Eng. Geosci.* 1 (1), 11–27.

Chapter 12

Disaster Mitigation Policies and Practices in South Asia with Focus on Social Capital

Jayant K. Routray and Saswata Sanyal

**Disaster Preparedness, Mitigation and Management (DPMM),
Interdisciplinary Academic Program,
Asian Institute of Technology,
PO Box 4, Klong Luang, Pathumthani, 12120, Thailand
E-mail: routray@ait.asia routray53@gmail.com, saswatasanyal@gmail.com**

ABSTRACT

Evidences prove that earth's climate has been changing at an unprecedented rate. South Asia is one of the most vulnerable areas towards the impacts of climate change. Decreased trend of availability of water in terms of quantity and quality, increase incidence of waterborne diseases, decrease in agricultural productivity and fishery production, adverse effects on ecosystems, increased risk of floods and droughts and increased damages and deaths caused by extreme weather events are some of consequences of climate change in this region. Furthermore, it has been well documented that over the years countries in South Asia have been severely affected by extreme climate events with increasing frequency and intensity. Two thirds of such disasters are induced by climate change in South Asia. The losses due to these climate-induced disasters are also huge. The magnitude of losses in this region gives an indication towards the importance of enhancing resilience against the extreme climatic hazards in order to ensure the steady path of economic growth and development in South Asia. In the context of extreme climatic events such as cyclone, flood, sea surge, and drought; most of the countries in South Asia have some policies that govern the various line departments and non-governmental institutions to undertake activities related to disaster risk reduction and mitigation. There is much more to be done for disaster mitigation than just limiting to structural or engineering measures like using non-structural measures.

Communities facing the challenges utilize the internal resources such as social capital. Social capital has a vital role to play in disaster mitigation, by helping the community to cope with stress and helps to mitigate adverse effects of a hazards, by using the bonding between community members and the connections they have with people outside their community. Therefore, it offers a great opportunity if the role and importance of social capital is clearly understood by the policy makers, it can be used to a great extent to initiate disaster mitigation strategies at the local level in the future.

Keywords: *Climate change, Extreme events, Climate-induced disasters, Disaster risk reduction, Disaster Mitigation, Policies and practices, Social capital.*

Introduction

Climate is changing at an unprecedented rate across the globe and there are a lot of evidences to ascertain it like, warming to climate at an unequivocal rate. According to Intergovernmental Panel on Climate Change, climate change is defined as any change in climate over time, whether due to natural variability or as a result of human activity (IPCC, 2014). Climate change is a natural phenomenon but has been substantially accelerated by anthropogenic activities in the recent history (O'Brien, O'Keefe, Rose, and Wisner, 2006). Emission of greenhouse gases by human beings is the single largest contributor of global warming across the globe along with degradation of the environment across the globe. (Helmer and Hilhorst, 2006)

Greenhouse gases (GHGs) in the atmosphere since the industrial revolution has increased substantially. The emission of GHGs like CO_2 (carbon dioxide) CH_4 (methane) N_2O (Nitrous Oxide) and several others has increased the greenhouse effect and is the main reason behind Earth's surface temperature to rise, thus cause Global Warming (EPA, 2016).Although industrialization and development did not happen uniformly across the world but global warming is a truly global phenomenon.

Climate change impacts the physical and biological systems in various ways (IPCC, 2007). Although the impacts of climate change are global, the actual changes and the risks can differ strongly from region to region. Effects of climate change and the risks associated with it is highest in the least developed countries, where they are most poorly equipped to deal with these adversities (O'Brien, O'Keefe, Rose, and Wisner, 2006). According to the IPCC:

Populations are highly variable in their endowments and the developing countries, particularly the least developed countries. have lesser capacity to adapt and are more vulnerable to climate change damages, just as they are more vulnerable to other stresses. This condition is most extreme among the poorest people (IPCC, 2001).

The effects of climate change are multifaceted and multidimensional. According to the 'Special Report on Emission Scenarios' (SRES) scenarios for greenhouse gas emissions the projections of temperature for the end of this century range from 1.1 to 6.4 °C when it is compared to end-20th century which leads to a wide range of effects on global, regional and local levels. These effects are changes (average and extremes) in temperature, sea levels, precipitation, food production, ecosystem health, species distributions, human health and extreme weather events (IPCC, 2007).Under extreme weather events following is happening and also expected to take place in the future (Pollner, Kryspin-Watson, and Nieuwejaar, 2008):

☆ Increase in temperature and decrease in mean precipitation leads to an increase in the frequency and severity of drought and heat waves.

☆ Increasingly warm ocean surface temperature generates more and stronger hurricanes, as well as commensurate flooding in the aftermath.

☆ Severe drought leads to an increase in forest fires.

☆ Greater intensity of wind and rain causes severe floods and landslides

Extreme weather events have been on a rise in recent years. Along with this rise there has been exponentially increasing economic losses, coupled with an increase in causalities due to these events, which has focused the attention of the world towards these events (Easterling, Evans, Groisman, Karl, Kunkel, and Ambenje, 1999).

South Asia, the home for one-fifth of the world's population, has been one of the most disaster-prone regions in the world. Decreased trend of availability of water in terms of quantity and quality, increase incidence of waterborne diseases, decrease in agricultural productivity and fishery production, adverse effects on ecosystems, increased risk of floods and droughts and increased damages and deaths caused by extreme weather events are some of consequences of climate change in this region (World Bank, 2009). In the region about two thirds of all the disasters the experiences are related to climate. There has also been a phenomenal increase in the frequency, severity and unpredictability of these climate related disaster in the recent times in this region. Between 1995 and 2015, there is account for more than 700 occurrences of disaster events that are induced by climate change resulting in more than 100,000 causalities, affecting more than 1 billion people and an economic loss of around $ 92 billion (EM-DAT, 2016).

Some of the impacts have been envisaged in the near future are increase in sea level rise leading to submergence of low-lying coastal areas and depletion of glaciers in the Himalayas threatening the perennial rivers that sustained food, water, energy and environment security of the region. The climate change also is unquestionably creating chances for more severe and newer risks of disasters in the region in the coming years (O'Brien, O'Keefe, Rose, and Wisner, 2006).Adding to these risk are the layers of vulnerabilities in this region which are poverty, illiteracy, malnutrition and social inequities, creating recipes of more disasters (World Bank, 2009).

The magnitude of losses, vulnerabilities of the people and risk in this region gives an indication towards the importance of enhancing resilience against the extreme climatic hazards in order to ensure the steady path of economic growth and development in South Asia. More people resettle in the hazardous coastal region particularly in urban areas in an unsafe and unplanned manner, prone to high risk of climate-induced hazards (Memon, 2012). The investments in South Asia for effective risk management warrant top priority and highest attention. Formulation of risk reduction and management policies and translating action in risk prone deserves equal priority vis-a-vis investment in risk reduction measures (World Bank, 2009).Policies related to disasters generally follow a top down approach in this region and are generally more focused on the physical aspects of disaster risk management. *But as much important the physical aspect of disaster risk management is, social aspects are also as important, considering that the social aspects are one of the most important links in managing disasters at the community level.*

In last decade, there has been a lot of discussion on social capital, which is one of the important resources embedded in the social structures of a community. Social capital very simply can be defined as 'the invisible resource that is created when people cooperate' (Coleman, 1990).

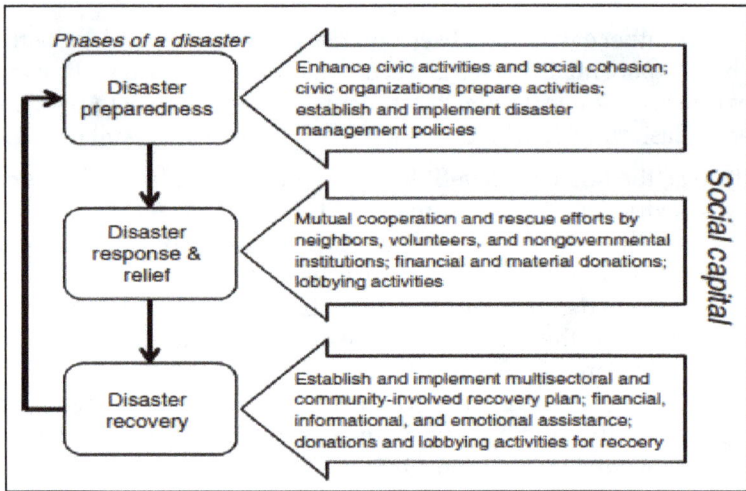

Figure 12.1: Links of Social Capital to Various Phases of Disaster (Aida *et al.*, 2013).

The above figure better explains the links of social capital to various phases of disaster (Aida *et al.*, 2013). From the figure it can be comprehended that social capital has many roles to play in disaster risk management. In the upcoming sections this paper will delve deeper into the policies in South Asia dealing with disaster and how social capital can benefit in disaster mitigation.

Materials and Methods

Journal papers, articles, websites, and reports were referred to, using the given keywords: Climate change, extreme events, climate-induced disasters, disaster risk reduction, disaster mitigation policies and practices and social capital. Searching of the documents were undertaken after realizing the aim of the paper, which was to understand the key issues of Disaster Risk Management in South Asia, to identify the gaps and to suggest measures to overcome the gaps in the future using social capital as a tool. After compiling papers, articles and reports, a thorough review was conducted in order to get a deeper understanding about the aim followed by summarizing of the evidences, and lastly interpreting the findings to arrive at the conclusion.

Results and Discussions

This section has two parts. First part covers the policies in South Asia for Disaster Risk Management whereas the second part discusses how Social Capital can contribute towards mitigation of disasters.

Policies in South Asian for Disaster Mitigation

The ever Increasing trends of natural disasters in South Asia and their threatening impacts on lives and livelihoods have resulted in a paradigm shift in disaster management in all the countries in the region from a post disaster relief and rehabilitation to holistic management of management of disasters covering all phases of disasters. The focus is clearly on Disaster Risk Management (DRM) - preparedness, mitigation and prevention (SAARC, 2008). This section will try to summarize initiatives taken by the governments in the region towards effective Disaster Risk Management.

Afghanistan

Institutional framework for disaster management was put in place in early 1980s by the Department of Disaster Preparedness (DDP). DDP was at the core, serving the Central Disaster Coordination Agency and secretariat of National Commission for Emergency Response (NCER). In 2007, DDP developed a strategy to establish an effective system of preparedness and response by 2010. The DRM institutional framework was revised in 2008. It is reflected in the National Disaster Management Law. Disaster risk reduction was also made as a priority goal for the Afghanistan National Development Strategy (ANDS) in 2008. After the revision of framework Afghanistan National Disaster Management Authority (ANDMA) came into being which coordinates and manages all aspects dealing with managing disasters with National Disaster Management Commission (NDMC) as the apex body (Memon, 2012; ANDMA, 2016).

National emergency fund (NEF) was also established to mobilize funds by NDMC for disaster related activities. National Emergency Operation Center (NEOC) looks after disaster preparedness and response activities based on National Plan for Disaster Management (NPDM). For decentralization functional offices have been setup by ANDMA within all provinces of the country. The Provincial Disaster Management Commission headed by provincial governors receive support from Provincial Disaster Management Agencies (PDMA). Further at the district level District Development Committees (DDC) and Community Development Councils (CDC) have been established all across the country (GFDRR, 2012; ANDMA, 2016). The 'Strategic Position on DRM' document also mentions Community Based Disaster Risk Management (CBDRM) as vital for building community resilience against disaster and climate shocks.

Despite the recent developments, the existing DRM structures and capacities remain weak partly owing to erosion of capacity due to decades of war and civil conflict and the country still heavily depends of outside aid for disasters. Government of Afghanistan (GoA) suffers from weak education system, non existence of early warning systems, lack of comprehensive risk assessments and poor national local linkage which execrates community vulnerability (GFDRR, 2012).

Bangladesh

Government of Bangladesh has put into place efficient response oriented infrastructure for managing disasters that was proven successful during Cyclone Sidr in 2007. Government of Bangladesh (GoB) has organized its DRM structure to proactively reduce risk of extreme events. National Disaster Management Council (NDMC) is the apex body which formulates and reviews disaster management policies. These policies of NDMC are implemented by Inter-Ministerial Disaster Management Coordination Committee with assistance from National Disaster Management Advisory Committee. Ministry of Disaster Management and Relief coordinates disaster preparedness and mitigation activities across all agencies. Bangladesh DRM strategy is detailed in National Plan for Disaster Management which provides a vision and policy direction for period between 2010-2015 with goal of integration DRM into development plans. Comprehensive Disaster Management Program (CDMP) has proved as a key DRM implementation program for Bangladesh by improving DRM capacity. It has been recognized for its work on community based risk assessments and risk reduction plans for flood and cyclones (Memon, 2012; DDM, 2016).

In spite of considerable efforts overall capacity remains weak thereby limiting the ability to carry out its mandate. Another vital shortcoming has been neglect of seismic risks as well as lack of sufficiently addressing threat from changing climate and its impacts. The system suffers from lack of strategies and mechanisms for effective disaster risk financing (GFDRR, 2012).

Bhutan

A vision of holistic DRM was established in DRM Framework Paper of 2006. Bhutan's 10th five-year plan has highlighted the importance of integrating Disaster Risk Management into development planning with focus on glacial lake outburst floods (GLOF) and earthquakes. Ministry of Home and Cultural Affairs serves as the focal agency for coordinating DRM activities at different administrative levels. The Department of Disaster Management (DDM) established in 2008 leads all DRM activities in the country and also functions as the national coordinating agency for disaster management. At district level Disaster Management Committees have been setup headed by the governor for performing disaster management related functions. Funding mechanism as well as disaster management activities for preparedness and mitigation have been envisaged in National Disaster Management Bill 2012 which was then passed as an act in 2013. Moreover, it also reflects the need for incorporating seismic resistant techniques into traditional forms of construction (Memon, 2012; DDM, 2016). The DDM has also understood the importance of building capacities at the local level and has been training its personnel at the local level on CBDRM (ADRC, 2009).

Despite recent progress in institutionalizing a proactive DRM strategy, many challenges remain. The current system at national and subnational level lacks DRM expertise technical expertise for hazard assessment and monitoring operations (GFDRR, 2012).

India

In the recent past large scale disasters like the Orissa Super Cyclone in 1999, Gujarat Earthquake in 2001 and the Indian Ocean Tsunami 2004 have prompted the government to have a proactive approach to disaster risk management in India. In 2005 the India's Disaster Management Act (DMA) was approved. It prescribes the establishment of a culture of disaster risk reduction, to be implemented at the national, state and local level. In each level, the development planning is required to integrate prevention and mitigation of disasters in the respective general and sectoral development plans, and also making provisions for resources to be directed towards mitigation activities. In 2005 DMA also established the National Disaster Management Authority (NDMA), which is chaired by the Prime Minister of India and is housed under the Ministry of Home Affairs, is responsible for making national level policy related to disaster management in India, providing guidelines and is responsible for all types of natural disasters except droughts. At the state level State Disaster Management Authority (SDMA) is mandated to function in a similar manner as does NDMA at the national level. DMA also established National Institute of Disaster Management (NIDM) for planning and promoting training and research in disaster management and the National Disaster Response Force (NDRF) for rapid response aftermath of a disaster (NDMA, 2016; Memon, 2012).

In India's 11th five-year plan there were provisions kept for developing a more holistic and integrated approach to DRM by transitioning from the relief and recovery approach to prevention, mitigation and preparedness approach. This was made into the National Policy on Disaster Management in 2009. The themes underpinning this approach according to the policy are, community based disaster management (including last mile integration of policy, plans and execution), capacity development in all spheres, consolidation of past initiatives and best practices, cooperation with agencies at National and International level and multi-sectoral synergy. Therefore, India has been successful in implementing a number of disaster preparedness programs across the country, introducing risk reduction initiatives, beginning building capacity for national risk mitigation programs and community based disaster risk management initiatives.

DRM in India has been improving because of all the institutions and the act. But despite these significant progresses at the national level and some state there has been limited progress in the case of DRM in most other states (GFDRR, 2012).

Maldives

The 7th five-year development strategy sets out DRM priorities to protect environment and livelihood including mainstreaming of disaster policies into respective line ministries in addition it also sets priority for improving public awareness through proper planning, preparedness and response. Indian ocean tsunami of 2004 lead to development of institutional framework for DRM in Maldives such as creation of National Disaster Management Center (NDMC) under Ministry of Defense and National Security (MDNS) which is responsible for overall coordination of disaster management activities, relief distribution and reconstruction (GFDRR, 2012).

DM act of 2006 lead to establishment of National Disaster Management Centre to serve as the policy making body. The act defines different responsibilities for NDMC with representation at the Central, Atoll and at the island level. It also provides a budget but leaves it to the treasury to establish National Disaster Response Fund (NDRF) for emergency assistance. National Disaster Management Center has helped in construction of multi-purpose community shelter and early warning systems. National Building codes have been introduced and community awareness programs have been launched to advance disaster preparedness efforts (Memon, 2012; NDMC, 2016).

Although considerable efforts have been put in, conditions still need to be improved. Insufficient human and financial resources accompanied by policy frameworks have slowed the implementation (GFDRR, 2012).

Nepal

Nepal, since 1980s has been actively involved in DRM and has long established institutions, policies and implementation capacity. The Natural Calamity (Relief) Act defines the current national DRM structure in Nepal. This acts specifically defines everyone's role in the government to engage in DRM activities as well as providing administrative structure for various DRM themes. Central Disaster Relief Committee (CDRC) is the apex institution presided over by the Ministry of Home Affairs (MoHA) and comprising 27 ministry secretaries and members of organisations (Police, Army, Red Cross, Meteorology Department) that collaborate for DRM (Memon, 2012; GFDRR, 2012).

According to the act in the sub-national level there needs to be establishment of regional committees on an ad-hoc basis along with permanent district level committees. In the past a number of these activities have been proven effective in managing relief after disasters. The District Disaster Relief Committee (DDRC) is headed by the Chief District Officer, and includes the district-level sectoral representatives (water, health, and education sectors). CDRC manages Central Disaster Relief Fund (CDRF) to fund DRM activities and it can be supplemented by the Prime Minister's fund. Since 2008 Nepal began to shift focus from an ex-poste disaster response approach to ex-ante disaster risk mitigation activities based on the existing and functioning DRM structures. In 2008, Government of Nepal (GoN) passed a National Strategy for Disaster Risk Management (NSDRM). The NSDRM operates as a guide for the transition towards a holistic approach to DRM, which includes focus on risk identification, vulnerability reduction and improved preparedness and response capacity. Following this strategy, the GoN has embarked on the formation of new institutional, legislative and policy frameworks for DRM (GFDRR, 2012). In the NSDRM, there is due importance given to the Community Based Organizations (CBOs) in order to promote community level disaster risk management activities. It has stressed on enhancing the existing capacities of the CBOs for DRM.

Although there has been a change in the stance of the government towards a more proactive approach towards DRM there are still some limitations in the existing capacities and institutions that prohibit them to realize their full potential and the also the enforcement of the legislation is lacking. In general, there is a lack of awareness of the risks associated with natural disasters and also the possible response mechanism both among the public and government officials (GFDRR, 2012).

Pakistan

In the wake of 2005 earthquake, the Government of Pakistan (GoP) made concerted efforts towards establishing a comprehensive disaster management regime by adopting a shift from reactive to a proactive approach. In 2006, the National Disaster Management Ordinance (NDMO) replaced the 1958 Calamity Act. In 2010, the National Disaster Management Act was passed to counter the threats of disaster faced by the country. In 2013, National Disaster Risk Reduction Policy (NDRRP) replaced the NDMO. According to the NDRRP, disaster risk reduction interventions were being carried out in Pakistan by different agencies in isolation at the national, provincial and district levels. There was strong need to give all these agencies and departments some direction and guidelines at align their activities with the National Disaster Management Act. The GoP believes NDRRP can give a strong sense of direction to all these departments and agencies in Pakistan and help in achieving disaster risk reduction in a more holistic way by laying special emphasis on risk assessment, prevention, mitigation and preparedness. The National Disaster Management Authority (NDMA), which is the national coordinating agency for disaster risk reduction and, together with the Earthquake Reconstruction and Rehabilitation Authority (ERRA), is responsible for all aspects of DRM. At the subnational levels, the Provincial Disaster Management Authorities (PDMAs) and District Disaster Management Authorities (DDMAs) are mandated with undertaking disaster management functions (GFDRR, 2012; Memon, 2012).

Recent developments include the creation of a Ministry of Climate Change, to coordinate all disaster- related agencies and activities, and passage of the 18th Constitutional Amendment by Pakistan's Parliament, which devolved some powers to the provinces, including overall responsibility to prepare for and respond to disaster. According to the NDRRP, it is observed that at the community level risk awareness is usually higher in those areas that have been affected by disasters recently and subsequently involved in Community Based Disaster Risk Management (CBDRM). It is understood in the NDRRP that CBDRM has a pivotal role to play in strengthening the overall disaster risk reduction system. In Pakistan, national and international NGOs have largely been implementing CBDRM programs. But in absence of CBDRM framework, it becomes difficult to create synergies and get maximum benefits for hazard prone communities out of CBDRM efforts (GoP, 2013).

Significant progress in implementing the DRM agenda has been inhibited mostly due to lack of capacity within the government at the sub-national levels owing to insufficient resources available for disaster risk agenda, lack of trained professional staff at the disaster management authorities at all levels. Limited progress in undertaking hazard risk assessments and sharing of risk data has led to need for further mitigation interventions (GFDRR, 2012).

Sri Lanka

The Indian Ocean Tsunami of 2004 set the island country of Sri Lanka on a track to be more involved in DRM activities. The disaster management framework in Sri Lanka took shape in the aftermath of the 2004 tsunami and is now based on emergency preparedness and response. The Government of Sri Lanka (GoSL) after the tsunami has taken substantial efforts to reduce vulnerability towards adverse natural events which includes, strengthening the country's disaster monitoring and early warning systems, emergency preparedness and planning, increasing awareness and capacity of sub-national officials and schools, and, introducing and enforcing DRM aspects into land-use and development planning. A Select Committee was established by the Sri Lankan Parliament in 2005, which was given the responsibility to investigate the country's preparedness to meet emergencies. Based upon the recommendations of this select committee Sri Lanka Disaster Management Act came into being in 2005. The Act defines the roles of the National Council for Disaster Management (NCDM) as the apex body for DRM coordination and monitoring (Memon, 2012).

Disaster Management Center (DMC) is the executive agency responsible for implementing directives of NCDM, under the Ministry of Disaster Management. DMC is the national level nodal agency responsible for coordinating all aspects relating to DRM. It also promotes collaboration between local level DRM programs in order to ensure mainstreaming with sectoral development plans. Sri Lanka has been able to enhance its existing disaster management framework for preparedness and response including improved land-use planning, increased disaster awareness, conducting trainings and developing hazard maps (GFDRR, 2012). In 2013, GoSL also came out with the National Policy on Disaster Management. In the policy it was stated that the disaster prone communities have the right to participate and contribute in decision making processes related to DRR and response and for disaster risk assessment and management, ministries are encouraged to adopt community based consultative approaches. CBDRM is also accepted as a tool for disaster risk reduction at the local level. (GoSL, 2013)

While there has been much progress made since the 2004 tsunami, challenges still remain specially in terms of technical capacity to implement the national DRM plan. There is a need to enhance existing legal and policy provisions in order to enforce the national DRM plan (GFDRR, 2012).

Therefore, it is safe to say that in all countries of the region there has been a paradigm shift in the approach to deal with disasters from being reactive and relief based to being proactive and, preparedness and mitigation based. In all the countries the required acts and frameworks are present but in many of them there is a lack of technical expertise and capacity to carry out all the functions necessary for effective DRM. This is because of the lack of trained personnel. There is also a

Table 12.1: Summary of DRM in South Asia

Country	Apex Body	Coordinating Body	Sub-national Level Body	Act/Law	Gaps
Afghanistan	National Disaster Management Commission (NDMC)	Afghanistan National Disaster Management Authority	☆ Provincial Disaster Management Commission ☆ District Development Committee and Community Development Council	Afghanistan Disaster Management Law	☆ Weak Capacity and DRM structure ☆ Non-existent EWS ☆ Lack of comprehensive risk assessments ☆ Poor National-local linkage
Bangladesh	National Disaster Management Council (NDMC)	Inter-Ministerial Disaster Management Coordination Committee (IMDMCC)	☆ District Disaster Management Committee ☆ Upazila Disaster Management Committee (Sub-district level) ☆ Union Disaster Management Committee (Union level) ☆ Pourashava Disaster Management Committee (Municipality level) ☆ City Corporation Disaster Management Committee	Disaster Management Act, 2012	☆ Limitations in overall DRM capacities ☆ Neglect of seismic risks ☆ Lack of strategies and mechanism for effective disaster risk financing
Bhutan	Ministry of Home and Cultural Affairs (MoHCA)	Department of Disaster Management (DDM)	☆ Dzongkhag Disaster Management Committee ☆ Dungkhag Disaster Management Committee ☆ Thromde Disaster Management Committee ☆ Gewog Disaster Management Committee	Disaster Management Act of Bhutan, 2013	☆ Lack of technical expertise ☆ Lack of hazard assessment and monitoring operations
India	National Disaster Management Authority (NDMA)	National Disaster Management Authority (NDMA)	☆ State Disaster Management Authority ☆ District Disaster Management Authority	Disaster Management Act, 2005	☆ Limited progress in DRM in some states

Contd...

Table 12.1–*Contd...*

Country	Apex Body	Coordinating Body	Sub-national Level Body	Act/Law	Gaps
Maldives	National Disaster Management Center (NDMC)	National Disaster Management Center (NDMC)	☆ Atoll Disaster Management Authority ☆ Island Disaster Management Authority	Disaster Management Act, 2006	☆ Insufficient human resources ☆ Limited financial resources ☆ Slow implementation of policy frameworks
Nepal	Central Disaster Relief Committee (CDRC)	Central Disaster Relief Committee (CDRC)	☆ Regional Natural Disaster Relief Committee ☆ District Disaster Relief Committee	The Natural Calamity (Relief) Act	☆ Limitations in the capacities of DRM institutions ☆ Lack of enforcement of DRM strategy and legislation
Pakistan	National Disaster Management Authority (NDMA)	National Disaster Management Authority (NDMA)	☆ Provincial Disaster Management Authority ☆ District Disaster Management Authority	National Disaster Management Act, 2010	☆ Lack of capacity in the sub-national level ☆ Lack of technical human resource ☆ Limited progress in undertaking hazard and risk assessment
Sri Lanka	National Council for Disaster Management (NCDM)	National Council for Disaster Management (NCDM)	☆ District Disaster Management Coordinating Unit	Sri Lanka Disaster Management Act, 2005	☆ Limitations in technical capacity ☆ Need for enforcing DRM policy

huge problem in more than one country in enforcement of the acts or legislations. Although, in most of the countries at the national level their institutions and mechanism that are functional, it cannot be said at the sub-national and at the local levels. Other than this all the countries have a Command-and-control approach or a top-down approach towards disaster management, although it can be successful in channelizing resources from the national to the community level where the disaster strikes, there also needs to be community-based initiatives to actually understand better about the ground realities about risk and vulnerability and then act upon it.

Community Based Disaster Risk Management (CBDRM) has a prominent place in the national disaster management frameworks of all the countries in the region. Every country reflects through various documents that CBDRM is important and that it can help in empowering the communities, enhancing their capacities and involving them in every phase of disaster management including assessment of risks and preparation of plans for prevention mitigation, preparedness, response and recovery. This is thought to be done by integrating community structures and processes with local governing institutions. Moreover, because of lack of capacities at the local level in almost all countries in the region CBDRM become more important as a tool to reduce risk at the core. There are a lot of successful examples of CBDRM especially from Bangladesh and India, aided by INGOs and donor agencies. But there needs to be a solid framework for CBDRM in all countries which needs to be included in the policy documents. Therefore, a lot of work remains to be done in translating these commitments into action regarding CBDRM (SDMC, 2008).

Social Capital and Disaster Mitigation

Social capital refers to trust, norms and networks that affect social and economic activity within one community. High accumulation of social capital contributes significantly to social, political and economic performance. There are many different definitions of social capital, as it has developed over the last several years based upon the work of several authors. Some have argued on the positives of social capital while others have argued on the negative impacts (Nakagawa and Shaw, 2004).

With more research being done in this area certain categorizations of social capital emerged. Two among them were the most fundamental to understand social capital. According to the work of Woolcock(1998), social capital has three categories, also explained in the Figure 12.2:

- ☆ Bonding social capital: It is defined as the strong ties between immediate family members, neighbours, close friends, and business associates sharing similar demographic characteristics.

- ☆ Bridging social capital: It is the ties between people from different ethnic, geographical, and occupational backgrounds with similar economic status. These ties a bit weaker than the bonding social capital ties.

- ☆ Linking social capital: It is the connection between people at the community level and people in positions of influence in formal organisations like government, banks or the police.

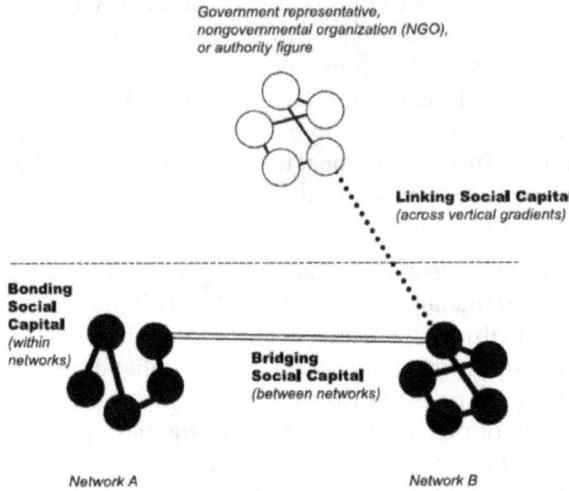

Figure 12.2: Diagram Depicting different Categories of Social Capital (Aldrich, 2012).

On the other hand, Uphoff (2000) categorized social capital into:

☆ *Structural social capital*: It includes roles, rules, precedents and procedures as well as networks that contribute to cooperation.

☆ *Cognitive social capital*: It is the mental process in human that is reinforced by culture, norms, values, attitudes and beliefs that contribute to cooperative behavior and mutually beneficial collective action.

It is difficult to choose the right definition of social capital as there are so many different views and perspective on the concept. So for this paper, we define social capital as *potential resource that stimulates multiple functions for mutual benefits carried out by members of the community.*

Social capital is vital for community resilience efforts for preparedness, response and recovery to environmental disasters. But unfortunately the potential role and contribution of local level social organizing enabled by social capital is overlooked by the policymakers (LaLone, 2012).

Dynes (2006) emphasizes that there needs to be more attention given towards social capital which can help the community to deal with a disaster, other than just focusing on damage on physical and human capital. Disaster management all across the globe generally follows command-and-control approach. Although, this approach is efficient when we look at channeling formalized emergency resources but its weakness is the assumption that the community either breaks down or has insignificant contributions to make in the different stages of a disaster. This makes the disaster mitigation and management policy makers to overlook or exclude the contribution of social capital from being factored into disaster management policies. Literature also shows the successes that have occurred when projects have adopted community-based participatory approach, where the social capital and its networks were intentionally identified and factored into inclusive partnership approach for community planning rather than overlooking their potentials (LaLone, 2012). Other than having structural mitigation measures it is also important to identify and use informal community networks for understanding local risks and needs, and thus make a more comprehensive disaster mitigation at the local level (UN-ISDR, 2005). Disaster mitigation helps in reducing and limiting the disruptive and destructive effects of hazards. As mentioned earlier disaster mitigation measures range from the physical such as engineering works like dikes, embankments, and safe building design to the non-structural measures such as community risk assessment, community risk reduction planning, public awareness, food security programs, group savings, cooperatives, crop insurance, strengthening the organizations for community disaster management and advocacy on disasters and development issues, legislation and land use zoning (Victoria, 2002).

It has been observed through practices that social capital networks when channelized can greatly help in reducing vulnerability through building capacities of communities to mitigate, prepare for and respond to disaster in a self-reliant and cooperative manner. Following are the steps by which social capital networks at the community level can be used for better preparedness and mitigation involved (LaLone, 2012; Victoria, 2002):

- ☆ Selecting project sites with the help of social networks and targeting the most vulnerable communities
- ☆ Selecting community members best suited to be volunteers and training them to work with communities in reducing vulnerabilities
- ☆ Organizing communities and establishing village-level Disaster Management Committees (DMC) as a coordinating body which can also plan ahead of disasters and emphasize community based-discussions
- ☆ With the help of the social network identifying, estimating and ranking local disaster risks through risk mapping
- ☆ Building consensus on mitigation solutions for better acceptance among the community members
- ☆ Using the social network to mobilize of resources and implementation of community mitigation solutions

Since every country in the region has a framework for DRM and most of the countries also have projects at the local level, it will be really beneficial to use social capital at the local level projects for better knowledge about the community and better involvement and ownership from the community members towards these projects. Other than this, while reviewing the policies and practices in the region it was observed that the DRM institutions at the local level have limited capacities, therefore if policies focus on strengthening the community using social capital, it can make the communities more resilient.

Conclusion

Disaster occurs at the community level. The community which is impacted by the disaster is a direct and active participant at all distinct phases of dealing with a disaster which are: preparedness, response, recovery and mitigation (Mushkatel and Weschleer, 1985).For the communities living in hazardous areas other than structural mitigation measures, it is important to develop non-structural mitigation measures also. Consideration of social aspects is one of the most important links in managing disasters at the community level.

There needs to be more integration of the factors that are in action at the community level which makes the community member more vulnerable or resilient when planning for disaster risk mitigation and management complimenting the structural measures. Thorough research has proven that Social Capital is one such resource which helps in increasing the resilience of the people if used properly (Aldrich, 2012).

Social capital has an important part to play in disaster mitigation, by assisting the community cope with stress and help to mitigate adverse effects of a hazards. It has been observed that individuals who participated in groups activities in their community had significantly mitigated their risk of succumbing to a disaster (Koh and Cadigan, 2008). Advantages that social capital poses should be taken into consideration by the policy makers while planning for disaster mitigation and preparedness in the future, especially when planning about community-based initiatives where social capital is inherent and can be mobilized with ease.

References

1. ADRC. (2009). *Bhutan DRR Policy Peer Review Report.* Retrieved from Asian Disaster Reduction Center: www.adrc.asia/publications/drr/pdf/Peer_Review_Bhutan.pdf

2. Aida, J., Kawachi, I., Subramanium, S. V., and Kondo, K. (2013). Disaster, Social Capital and Health. In *Global Perspective on Social Capital and Health* (pp. 167-187). New York: Springer.

3. Aldrich, D. P. (2012). *Building Resilience : Social Capital in Post-disaster Recovery.* Chicago: The University of Chicago Press.

4. ANDMA. (2016). *About ANDMA.* Retrieved from Afghanistan National Disaster Management Authority: http://andma.gov.af/#

5. Coleman, J. (1990). *Foundations of Social Theory.* Cambridge, MA: Belknap Press, Harvard University Press.

6. DDM. (2016). *About Us.* Retrieved from Department of Disaster Management: http://www.ddm.gov.bt/

7. DDM. (2016). *Department of Disaster Management.* Retrieved from Department of Disaster Management, Ministry of Disaster management and Relief: http://old.ddm.gov.bd/

8. Dynes, R. (2006). Social Capital: Dealing with Community Emergencies. *The Journal of the NPS Center For Homeland Defense and Security*.

9. Easterling, D. R., Evans, J. L., Groisman, P. Y., Karl, T. R., Kunkel, K. E., and Ambenje, P. (1999). Observed Variability and Trends in Extreme Climate Events: A Brief Review. *Bulletin of the American Meteorology Society*, 417-425.

10. EM-DAT. (2016, 2 24). *International Disaster Database*. Retrieved from Centre for Research on the Epidemiology of Disasters: www.emdat.be

11. EPA. (2016, 2 24). *Causes of Climate Change*. Retrieved 2 24, 2016, from United States Environmental Protection Agency: http://www3.epa.gov/climatechange/science/causes.html

12. GFDRR. (2012). *Disaster Risk Management in South Asia: A Regional Overview*. The World Bank, Global Facility for Disaster Reduction and Recovery. Washington D.C.: The World Bank.

13. GoP. (2013). *National Disaster Risk Reduction Policy*. Retrieved from NDMA: http://www.ndma.gov.pk/Documents/drrpolicy2013.pdf

14. GoSL. (2013). *National Policy on Disaster Management*. Retrieved from Ministry of Disaster Management: http://www.disastermin.gov.lk/web/images/pdf/draft per cent 20dm per cent 20policy.pdf

15. Helmer, M., and Hilhorst, D. (2006). Natural Disasters and climate change. *Disasters*, 30 (1), 1-4.

16. IPCC. (2001). *Climate Change 2001: Working Group II: Impacts, Adaptation and Vulnerability. Summary for Policymakers*. Geneva: Intergovernmental Panel on Climate Change.

17. IPCC. (2007). *Climate change 2007: Impacts, adaptation and vulnerability. Contribution of working group II to the Fourth Assessment Report of the Intergovernmental Panel on Climate Change*. New York: Intergovernmental Panel on Climate Change (IPCC), Cambridge University Press.

18. IPCC. (2014). *Climate Change 2014: Synthesis Report. Contribution of Working Groups I, II and III to Fifth Assessment Report of the Intergovernmental Panel of Climate Change*. Geneva: IPCC.

19. Koh, H. K., and Cadigan, R. (2008). Disaster Preparedness and Social Capital. In *Social Capital and Health* (pp. 273-285). Springer.

20. LaLone, M. B. (2012). Neighbors Helping Neighbors: An Examination of the Social Capital Mobilization Process for Community Resilience to Environmental Disasters. *Journal of Applied Social Sciences*, 6 (2), 209-237.

21. Memon, N. (2012). *Disasters in South Asia*. Karachi: Pakistan Institute of Labour Education and Research.

22. Mushkatel, A. H., and Weschleer, L. F. (1985). Emergency Management and the Intergovernmental System. *Public Administration Review*, 45 (Jan), 49-56.

23. Nakagawa, Y., and Shaw, R. (2004). Social Capital: A Missing Link to Disaster Recovery. *International Journal of Mass Emergencies and Disasters*, 5-34.

24. NDMA. (2016). *About National Disaster Management Authority*. Retrieved from National Disaster Management Authority: http://www.ndma.gov.in/en/

25. NDMC. (2016). *About NDMC*. Retrieved from National Disaster Management Centre: http://ndmc.gov.mv/wordpress/?page_id=22

26. O'Brien, G., O'Keefe, P., Rose, J., and Wisner, B. (2006). Climate change and disaster management. *Disasters, 30(1)*, 64-80.

27. Pollner, J., Kryspin-Watson, J., and Nieuwejaar, S. (2008). *Disaster Risk Management and Climate Change Adaptation in Europe and Central Asia*. The World Bank, Global Facility for Disaster Reduction and Recovery. GFDRR.

28. SAARC. (2008). *Regional Cooperation on Climate Change Adaption and Disaster Risk Reduction in South ASia*. South Asian Association for Regional Cooperation. Kathmandu: SAARC.

29. SDMC. (2008). *Community Based Disaster Risk Management in South Asia*. SAARC Disaster Management Center. New Delhi: SAARC Disaster Management Center.

30. UN-ISDR. (2005). *Hyogo Framework for Action 2005-2015*. Geneva: United Nations Internation Strategy for Disaster Reduction.

31. Uphoff, N. (2000). Understanding Social Capital: Learning from the Analysis and Experience of Participation. In P. Dasgupta, and I. Serageldin (Eds.), *Social Capital: A Multifacted Perspective* (pp. 215-249). Washington D.C.

32. Victoria, L. P. (2002). Community-based Approaches to Disaster Mitigation. *Regional Workshop on Best Practices in Disaster Mitigation Lessons Learned from the Asian Urban Disaster Mitigation Program and other initiatives*. Bali: Asian Urban Disaster Mitigation Program.

33. Whittaker, H. (1979). *Comprehensive emergency management: A Governor's guide*. National Governors' Association. Washington, DC: Center for Policy Research.

34. Woolcock, M. (1998). Social capital and economic development: Towards a theoretical synthesis and policy framework. *Theory and Society* (27), 151-208.

35. Woolcock, M. (2001). The place of social capital in understanding social and economic outcomes. *Canadian Journal of Policy Research, 2* (1), 11-17.

36. World Bank. (2009). *South Asia Climate Change Strategy*. NA: The World Bank.

Chapter 13

Existing National Policies on Natural Disaster Management and Implementation: A Case of Landslides and Mudslides in Uganda

Kisamba Mugerwa

Executive Chairperson,
National Planning Authority
Planning House, Plot 15B Clement Hill Road
P.O.Box 21434, Kampala-Uganda
E-mail: wkisambamugerwa@npa.ug, wkisambamugerwa@gmail.com

ABSTRACT

This paper addresses the existing national policies on natural disaster management and implementation with particular emphasis on landslides and mudslides experienced in Mt. Elgon areas in eastern Uganda. It specifically addresses the institutional arrangement for disaster mitigation efforts, the challenges faced in implementing the policy in relation to landslides and mudslides, and the suggestions for improving the existing policy of disaster measures relevant to landslides and mudslides. The concern of this paper is that disasters and emergencies are too often regarded as abnormal events, divorced from normal life. The magnitude of each disaster, measured in terms of deaths, damages to lives and property, or costs both real and monetary, depends on the level of vulnerability of the population. Vulnerability to landslides and mudslides in Uganda is high because of the country's heavy reliance on climate dependent resources and activities such as rain-fed agriculture, degraded ecosystems, and weak institutional capacity. Indeed a Climate risk report (CIGI 2007) labeled Uganda as one of the most unprepared and most vulnerable countries in the world. More fundamentally however is the realization that severe climatic events are increasingly causing far reaching disasters in Uganda. This manifestation of disasters comes in various forms. Yet, mitigation efforts and disaster planning remain elusive despite the creation of disaster management institutions by the government and the increased awareness by civil society and researchers, technical and environmental practitioners of the grave implications that severe landslides and mudslides events impose.

Keywords: Climate, Vulnerability, Disaster, Mitigation, Policy, Mudslides, Landslides.

Introduction

Disasters and emergencies are too often regarded as abnormal events in life. Sometimes they are consequences of the ways societies structure themselves economically and socially and how society interacts with the environment. Factors responsible for occurrence of disasters are basically geological, meteorological and social. The magnitude of each disaster are measured in terms of deaths, damages to lives and property, or costs both real and monetary, depends on the level of vulnerability of the population.

Vulnerability to climate change in Uganda is high because of the country's heavy reliance on climate dependent resources and activities such as rain-fed agriculture, degraded ecosystems, weak institutional capacity, limited infrastructure and technology, poor human resources and equipment for disaster management, limited financial resources and widespread poverty. Indeed a Climate Risk Report (CIGI, 2007) labeled Uganda as one of the most unprepared and most vulnerable countries in the world.

The demand of the hour is to work together to find mitigation options so that no further damage is done, while where necessary, adapt to the changing climate. The Intergovernmental Panel on Climate Change(IPCC, 2015) defines mitigation as a technological change and substitution that reduce resource inputs and emissions per unit of output with respect to climate change. Mitigation means implementing policies and technological measures. Measures can include affordable safety practices such as shelters, devices and equipment; early warning systems; awareness programs; scientific and sociological methods and tools in reduction of risk.

While adaptation aims to lessen the adverse impacts of climate change through a wide-range of system-specific actions (Fussel and Klein, 2002), mitigation looks at limiting risks. In this respect, mitigation looks at limiting risks related to landslides and mudslides.

More fundamentally however is the realization that severe climatic events are increasingly causing far reaching disasters in Uganda. This manifestation of disasters comes in various forms. Droughts, high temperatures, global warming, floods, mudslides and landslides are exerting unimaginable repercussions on people's lives. Yet, mitigation efforts and disaster planning remain elusive despite the creation of disaster management institutions by the government and the increased awareness by civil society and researchers, technical and environmental practitioners of the grave implications that severe climate events impose.

Against this background, this paper particularly addresses the mitigation efforts of landslides and mudslides in Mt. Elgon areas in eastern Uganda through an examination of the existing policy of disaster preparedness and management. It specifically addresses the institutional arrangement for disaster mitigation efforts, the challenges faced in implementing the policy in relation to landslides and mudslides, and the suggestions for improving the existing policy of disaster preparedness and management.

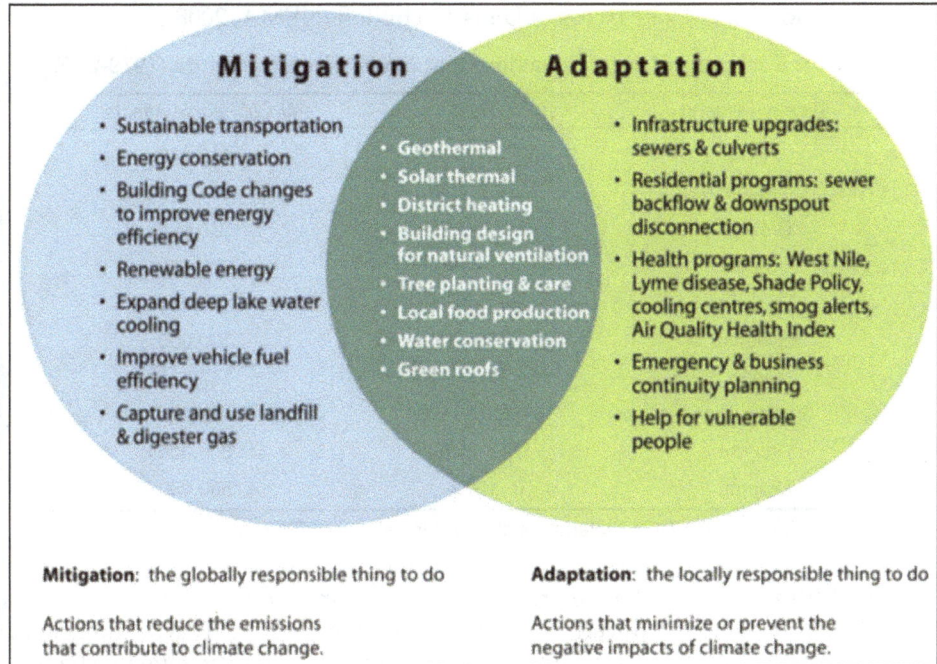

Mitigation

- Sustainable transportation
- Energy conservation
- Building Code changes to improve energy efficiency
- Renewable energy
- Expand deep lake water cooling
- Improve vehicle fuel efficiency
- Capture and use landfill & digester gas

- Geothermal
- Solar thermal
- District heating
- Building design for natural ventilation
- Tree planting & care
- Local food production
- Water conservation
- Green roofs

Adaptation

- Infrastructure upgrades: sewers & culverts
- Residential programs: sewer backflow & downspout disconnection
- Health programs: West Nile, Lyme disease, Shade Policy, cooling centres, smog alerts, Air Quality Health Index
- Emergency & business continuity planning
- Help for vulnerable people

Mitigation: the globally responsible thing to do

Actions that reduce the emissions that contribute to climate change.

Adaptation: the locally responsible thing to do

Actions that minimize or prevent the negative impacts of climate change.

Figure 13.1: Differences and Similarities between Adaptation and Mitigation.

Materials and Methods

This study was carried out through document analysis and consultations. Various documents were reviewed from various sources. These included technical reports, annual reports, technical reviews and working papers. Technical consultations were made from various organizations such Uganda Council for Science and Technology, Meteorological Centre Entebbe, Ministry of Disaster Preparedness, Office of the Prime Minister, Ministry of Finance, Planning and Economic Development. Consultations were also various technical people from Makerere University, National Planning Authority and Ministry of Agriculture, and Ministry of Lands and Water.

Results and Discussions

Main Natural Disasters in Uganda

Climate change induced disaster losses and damages are on the rise in Uganda in terms of scale, frequency and intensity with increasingly grave consequences for the survival and livelihood of the citizens, particularly the poor and vulnerable groups. Indeed, the country has been experiencing increased drought frequency and intensity and rising temperatures since 1960 and these trends are projected to continue in the future.

Climate related disasters in Uganda are estimated to contribute to over 70 per cent of natural disasters and destroy an average of 800,000 hectares of crops annually making economic losses in excess of UShs 120 billion (NEMA, 2008).

Table 13.1: Summary of some Disasters that have Affected Uganda (1979-2007)

Type of Disaster	Date	No. of People Affected
Flood	15 – Aug – 2007	718,045
Drought	Mar – 2005	600,000
Drought	June – 2002	655,000
Drought	Aug – 1999	700,000
Drought	Jan – 1998	126,000
Epidemic	26 – Nov – 1997	100,000
Flood	14 – Nov – 1997	153,500
Earthquake	6 – Feb – 1994	50,000
Drought	Dec – 1987	600,000
Drought	1979	500,000

Source: EM-DAT: The OFDA/CRED International Disaster Database, 2007.

In addition, climate change erodes Uganda's hard won development gains, frustrates efforts to reduce poverty and threatens human security. This phenomenon, compounded by environmental degradation, climate variability, climate change, and geological hazards, which if unchecked points to a future where disasters could increasingly undermine Uganda's economy and livelihoods of its people. For instance, in the past two decades, on average more than 200,000 Ugandans were affected every year by disasters (FAO, 2011). In 1987 alone, drought affected 600,000 people and related epidemic diseases killed 156 people two years later. In 1997 floods affected 153,500 people, killing 100. In 2005, drought affected 600,000 people and the following year (2006) climate related epidemic diseases killed 100 more. In the year 2007, floods affected 718,045 people while epidemic diseases killed 67 people and landslides 5 people (Uganda Bureau of Statistics, 2013). The above statistics clearly demonstrate the challenges posed by acute climatic events to the economic growth of Uganda and the security of its people.

In the last 100 years, Uganda's forests have faced severe pressures mainly from agricultural conversion as a result of population increase, urban demand for charcoal, over grazing, unplanned urbanization, uncontrolled timber harvesting and policy failures and inconsistencies. The forestry cover has shrunk from 45 per cent in 1890 to the current 14 per cent of the total land area with the rate of deforestation is estimated to be about 1 per cent per annum (Kigenyi, 2014). Deforestation, mainly the conversion of forests to agricultural land, continues at an alarming rate of approximately 13 million hectares per year (for the period 1990–2005, FAO).

Temperatures have increased by up to 1.5°C across much of Uganda with typical rates of warming around 0.2°C per decade, and will increase by up to 4.3 °C by the 2080 (Uganda Bureau of Statistics, 2013). Places such as Kabale and Fortpotal in western Uganda that used to be cold are no longer as cold. This transition to an even warmer climate is likely to amplify the impact of decreasing rainfall and periodic droughts, and will likely reduce crop harvests and pasture availability. The rise in temperature to cause a shift in the viability of coffee growing areas potentially wiping out US $ 265.8 million or 40 per cent of export revenue (DFID, 2008). This is set to exacerbate poverty and trigger migration as well as heightening competition over strategic water resources, indeed climate change could lead to regional insecurity.

The impacts of climate change are worsened by the high population growth. In 2011, Uganda's population was 30.6 million, with a rapid population growth rate of 3.6 per cent (CIA, 2011). If sustained, this growth rate will result in a doubling of the population every 20 years. According to Gridded Population of the World statistics (CIESIN, 2010), the population increased by 89 per cent between 1990 and 2010, adding some 15 million people.

On the other hand, the declining trend in lake levels especially Lake Victoria (2005-2008) from the mid-1960s to the present suggests a slow return to pre-1960 levels. The accelerated decline between 2004 and 2007 has been the subject of much concern and debate in the region (Sutcliffe and Peterson 2007). Lake Victoria is highly sensitive to changes in climate (DFID, 2008). The 1.6m drop in Lake Victoria level in 2005-2008 caused significant impacts on; hydropower generation dropping from 320 MW to 120 MW, disruption in urban water supply, fish landing sites, irrigation,

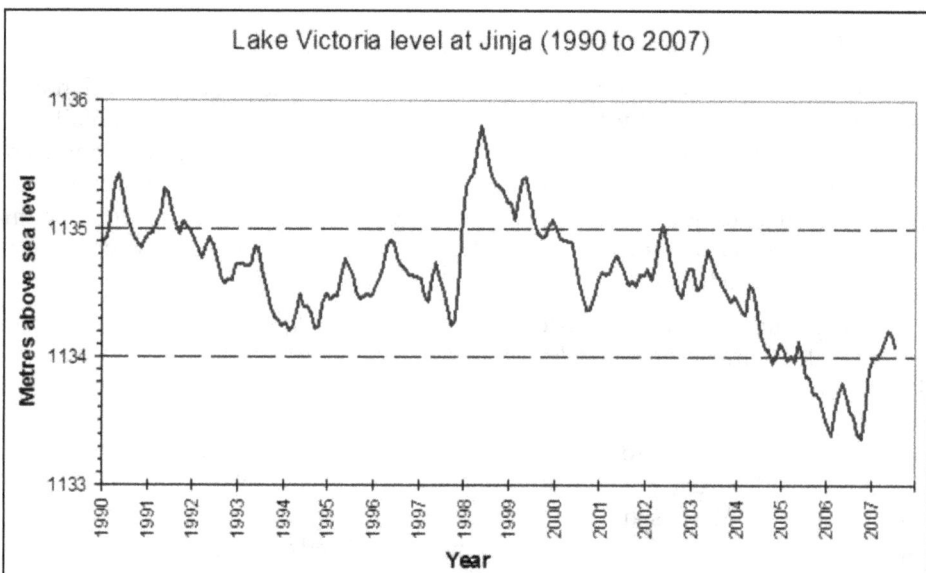

Figure 13.2: Lake Victoria Levels at Jinja between 1990 and 2007.

navigation and threats on the health of ecosystem. It also caused tensions between Uganda,Tanzania and Kenya.

Temperatures around the Lake had always varied but had increased consistently by 0.02-0.03°C annually since the 1980s, with 86 per cent of the loss through evaporation, resulting in a negative water balance and the failure of the lake to retain its historical water levels(Uganda Bureau of Statistics, 2013). In addition, climate change is reducing the size of several species of fish on lakes in Uganda and its neighboring East African countries, with a negative impact on the livelihoods of millions people who depend on fishing for food and income.

There is also an issue of precipitation as shown by observed rainfall reductions of the 1960–2009that are projected to the period 2010–2039. The projected rainfall declines range from -150 to -50 mm across the northern part of the country, and appear likely to impact the already chronically insecure IDPs and the inhabitants of Karamoja (Uganda Bureau of Statistics, 2013).On the other hand, higher mean and increased rainfall intensity raises risks of loss of life and property and damage infrastructure such as roads and bridges. Climate induced damages to infrastructure are currently $20-130 million a year, if there are no increase in frequency or intensity up to 2050, then damages rise to $39-234 million by 2025 and to $189-838 million by 2050 (Metroeconomica, *et al.*, 2015).

In Uganda, community settlement on steep slopes and other uncontrolled land use practices have increased the likelihood of landslides and mudslides prevalence. The areas mostly affected by Landslides are Mt. Elgon region, Ruwenzori region and Kigezi. Notable among the most affected areas is Mt. Elgon region more especially the densely populated districts surrounding the mountain slopes. These include Bududa, Manafa, Sironko and Bulambuli, which is a subject of our discussion in this paper.

Landslides and Mudslides

Landslides and mudslides are rapid movement of a large mass of mud, rocks, formed from loss soil and water by gravity. It usually follows heavy rainfall and high ground water flowing through cracked bed rocks and earth quakes and lead to movement of soils or sediments. Landslides and mudslides are very difficult to predict but their frequency and extent can be estimated by use of information on the area's geology, geomorphology, hydrology, climate and vegetation cover and traditional knowledge.

In Uganda, in Bududa district, in 1997 alone, 92 landslides have displaced about 11,000,000m³ of soil and debris into river channels and wetlands downstream (Kitutu, 2010). While in 2010, landslides killed about 250 people with over 8,500 affected. In 2012, heavy rains caused a serious landslide in Bulucheke Sub County, Bumwalukani Parish, Bududa District, burying the villages of Namanga and Bunakasala and affecting approximately 4,944 acres of land. The landslide claimed 3,480 persons in 421 households which were completely buried, while many others were severely damaged. The initial assessment identified 421 households with 3,480 persons affected by the disaster.

In the worst hit villages of Namanga and Bunakasala, large areas were washed away, leaving the farmers facing food shortages. In addition, numerous families lost their livestock, which put them in a vulnerable position since they lost their livelihood. This also hampered their recovery from the landslide. Several villages remained at risk of landslides as more cracks in the soil were being discovered.

The National Policy for Disaster Preparedness and Management

Paragraph XXIII of the National Objectives and Directive Principles of State Policy of the 1995 Constitution obliges the state to institute effective machinery for dealing with any hazard or disaster arising out of natural calamities or any situation resulting in the general displacement of people or general disruption of normal life. The National Policy for Disaster Preparedness and Management was enacted in 2010.The policy mission is to create an effective framework through which Disaster Preparedness and Management is entrenched in all aspects of development processes, focusing on saving lives, livelihoods and the country's resources. The goal of the policy is to establish institutions and mechanisms that will reduce the vulnerability of people, livestock, plants and wildlife to disasters in Uganda.

Some of the policy objectives are to establish Disaster Preparedness and Management institutions at national and local government levels; Equip Disaster Preparedness and Management institutions and ensure that the country is prepared at all times to cope with and manage disasters; Integrate Disaster Preparedness and Management into development processes at all levels; Promote research and technology in disaster risk reduction; Generate and disseminate information on early warning for disasters and hazard trend analysis, and; Create timely, coordinated and effective emergency responses at national, district and lower level local governments. This policy intends to complement the macro and sectoral policies currently being pursued by government. It takes into account the prevailing economic trends. in the country and utilizes a systematic, interdisciplinary approach which will ensure the integrated use of natural and social sciences in planning and decision making.

The main thrust of this policy is to make disaster management an integral part of the development process. It recognizes the profound impact of human activity on the interrelations within the natural environment as well as the influence of population growth, the high density of urbanization, industrial expansion, resource exploitation and technological advances. The policy also emphasizes the critical importance of restoring and maintaining the quality and overall welfare and development of human beings in their environment. A further fundamental purpose of the policy is to advocate an approach to disaster management that focuses on reducing risks – the risk of loss of life, economic loss and damage to property. This approach involves a shift from a perception that disasters are rare occurrences managed by emergency rescue and support services. A shared sense of awareness and responsibility needs to be created to reduce risks in our homes, communities, places of work and society in general.

Some of the guiding principles of the policy include sound planning using a multi-sectoral approach, community participation, public awareness and education institutional capacity building, adequate expertise and technology and vulnerability analysis.

Institutional Arrangement for Disaster Preparedness/Mitigation in Uganda

The implementation of the actions for national disaster management is a multi-sectoral and multidisciplinary process. The process involves all government ministries in collaboration with humanitarian and development partners, the private sector, local governments and the community. The Ministry responsible for Disaster Preparedness and Refugees in the Office of the Prime Minister is the lead agency in co-coordinating all stakeholders on disaster preparedness and management in the country.

Disaster preparedness and management is a shared responsibility between the state and all citizens. The overall goal of the institutional framework is to create and establish efficient institutional mechanisms for integrating disaster preparedness and management into the socio-economic development planning processes at national and local government levels. At the top of the institutional framework for disaster management is the President of the Republic of Uganda. Article 110 of the 1995 Constitution of Uganda gives the President the mandate to declare a state of emergency in any part of the country in the event of a disaster. The Minister in charge of Disaster Preparedness and Management provides the President with all the relevant details on the cause and effects of the disaster and mitigation and relief measures to be undertaken; in a situation where the disaster is caused by a natural or human-induced hazard. The President in consultation with Cabinet declares an area or the nation to be in a state of disaster.

The Cabinet is second in line. It the chief policy making body of government and advises the President on disaster related matters. It receives on quarterly basis a national food security report. It also issues policy direction on vulnerability management, hazard and risk mapping of the country and disaster risk reduction in general. There is also a Ministerial Policy Committee (MPC), which is a standing committee of Cabinet that handles cross sectoral matters relating to disaster preparedness and management. The committee ensures that disaster preparedness and management is mainstreamed in the governance of Uganda. The committee is chaired by the minister responsible for relief, disaster preparedness and refugees.

The Directorate of Relief, Disaster Preparedness and Refugees in Office of the Prime Minister (OPM) is the lead agency responsible for disaster preparedness and management. It shall coordinate risk reduction, prevention, preparedness, mitigation and response actions in the country in consultation with other line ministries, humanitarian and development partners, Local Governments and the Private sector. The Minister responsible for disaster preparedness and refugees links the Office of the Prime Minister to Cabinet.

Other concerned ministries include The Ministry for Agriculture, Animal Industry and Fisheries, the Ministry for Health, the Ministry for Water and Environment, the Ministry for Works and Transport, the Ministry for Housing and Urban Development, the Ministry for Housing and Urban Development, the

Ministry for Energy and Mineral Development, and the Ministry for Defense, the Ministry for Internal Affairs (Uganda Police Force), the Ministry for Information, the Ministry Responsible for Education. There is also an Inter-Agency Technical Committee that comprises of focal point technical officers from line ministries, UN agencies, NGOs and relevant stakeholders. In addition, there is a National Emergency Coordination and Operations Centre (NECOC)that deals with sudden-on-set emergencies such as mass casualty transport accidents, massive landslides and floods and collapsed buildings.

More so, there is a District Disaster Policy Committee (DDPC) whose functions are to give policy direction to the District Disaster Preparedness and Management Technical Committee; provides a link between national Disaster Preparedness and Management Committee and the Local government structures responsible for disaster preparedness and management. Local Communities and individual families are responsible for taking measures within their own capacities, to protect their own livelihoods and property. However, it is expected that measures taken by individual families and communities will for part of an integrated approach which will include the development of family management capacities and a reduction in their vulnerability over time.

Mitigation Efforts for Disaster Management

There is now international acknowledgement that efforts to reduce disaster risks must be systematically integrated into policies, plans and programs for sustainable development and poverty reduction. The Copenhagen Accord issued following the United Nations Climate Change Conference in Copenhagen (COP 15) in December 2009 provides that developing nations would implement mitigation actions (Nationally Appropriate Mitigation Actions) to slow growth in their carbon emissions.

The Government of Uganda, the Ministry of Water and Environment Climate Change department and in collaboration with UNDP launched a Low Emission Capacity Building (LECB) project. The LECB project focuses on strengthening Uganda's technical and institutional capacity in the development of national Greenhouse gas (GHG) inventory systems and Nationally Appropriate Mitigation Actions (NAMAs) and their associated measuring, reporting and verification systems. Key sectors identified in Uganda include agriculture, energy, and transport and waste.

The NAMA process illustrates that the Government of Uganda is living up to its international commitments under the United Nations Framework Convention on Climate Change (UNFCC) to combat global climate change. This was in realization, as mentioned above, of the effects of climate change in Uganda such as receding of ice caps on Mt. Rwenzori in western Uganda to about 60 per cent of their 1955 recorded area cover, temperature rise has resulted in malaria incidences, and rainfall has become erratic since 1990s; el Nino has resulted in over 1,500 deaths and over 150,000people displaced.

The NAMA process seeks to promote appropriate mitigation actions that are specifically in line with key Government policies and national priorities. These are the National Development Plan 2015/16-2020, the national Climate Change Policy 2013 and the Uganda Vision 2040. One of the major objectives of the National Development Plan is to promote sustainable use of environment and natural resources as well as good governance across sectors hence promoting sustainable development.

The overall goal is to ensure a harmonized and coordinated approach towards a climate resilient and low carbon development path for sustainable development in Uganda. This is done with an overarching objective to ensure that all stakeholders address climate change impacts and their causes through appropriate measures while promoting sustainable development and a green economy.

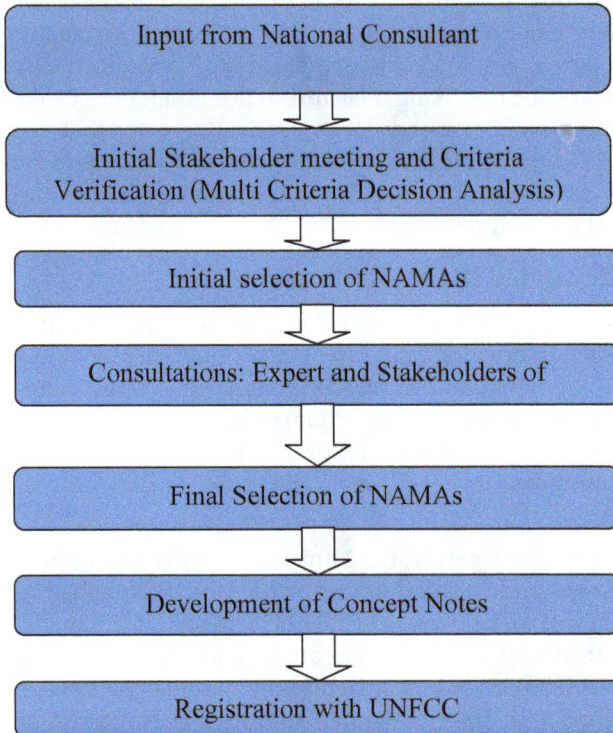

Input from National Consultant

Initial Stakeholder meeting and Criteria Verification (Multi Criteria Decision Analysis)

Initial selection of NAMAs

Consultations: Expert and Stakeholders of

Final Selection of NAMAs

Development of Concept Notes

Registration with UNFCC

Figure 13.2: The NAMA Selection Process in Uganda.

In line with the above, the specific policy actions targeting landslides and mudslides include:

(i) Gazette landslide and mudslide prone areas and strictly prohibit settlement in such risk areas;

(ii) Resettle all persons living in land/mudslide prone areas through creation of attractive resettlement packages;

(iii) Undertake measures to promote afforestation and reforestation;

(iv) Enforcement of relevant laws and policies;

(v) Apply appropriate farming technologies and land use practices;

(vi) Demarcation of land surrounding affected areas;

(vii) Undertake measures to reduce the high population growth rate through targeted investments including non-farm economic activities in the region and investment in human resource development.

More so, numerous discussions and engagements have been held with the landslides and mudslides affected districts' political and technical leadership involving Uganda Wildlife Authority, Office of the Attorney General, Ministry of Lands, Housing and Urban Development, Office of the Prime Minister, Office of the President, Ministry of Water and Environment, National Forestry Authority and the local communities.

The controversial issue surrounding the boundaries of Mt. Elgon was revisited to cover boundary reopening, provision of seedlings for tree planting in the affected areas, resettling the affected communities and undertake appropriate compensation. This will be accompanied by Uganda Wildlife Authority to exercise maximum restraint in enforcement of the law and use civic means of apprehending encroachers and other wildlife crime offenders in the area.

Discussions

There are several areas of concern regarding mitigation measures for disasters due to severe climate change. Though policies are available, they are not consciously implemented. Major areas of distress are centered on three issues: detection, prevention, and implementation. The extent of detection is undermined by capacity to detect severe climatic events in advance, technology, resources and personnel. Yet implementation is dented by poor coordination, collaboration and response speed.

Although, in 2010, the National Disaster Preparedness Policy was put in place in Uganda, with the overall policy goal of promoting national vulnerability assessment, risk mitigation, disaster prevention and preparedness, little has been implemented, with increasing encroachment on wetlands and gazetted areas that have continued to cause severe mudslides and landslides in Mt. Elgon areas. Inadequate plant seeds have also been provided for afforestation in areas affected by deforestation and landslides. Resettlement of people affected by climatic disasters has not been done in advance. Until now people in Bududa and Manafa areas of Mt. Elgon with severe mudslides have not been resettled. Land for resettlement has been grabbed.

There is also limited investment in research to ensure appropriate technologies to enable mitigation of mudslides and landslides in areas affected with mudslides and landslides. At the same time land use practices such afforestation ad mixed farming have not been utilized in areas affected by landslides and mudslides.

Conclusions

Demarcation and its enforcement still affect issues in disaster management in the areas of mudslides and landslides. Though demarcations are determined scientifically and should be based on proven data for areas that are more fragile, convincing the masses within these areas that they should vacate such areas is challenging. Enforcement is also very expensive in terms of staff and financial resources.

Poverty also stifles disaster management. Poverty leads to inadequacy of human needs for domestic use. Thus people encroach on reserved areas for firewood, building materials and food leaving the areas susceptible for landslides and mudslides.

Poor technology for agricultural practices subjects the affected areas to risks of mudslides and landslides. The technology used in rural areas is so rudimentary that they expose the soil to weather vagaries. Coordination is so weak at all levels of institutional arrangement that there is over departmentalization of different roles for disaster management. Instead of fostering synergies, departments are running a stand-alone silo style of handling disasters.

Disaster is a cross-cutting issue, multi-disciplinary and multi-sectoral. However, disaster management efforts also face the difficulty of people's mindsets to change. People are so ingrained in traditional farming practices as well as ways of living that are so dependent on subsistence survival, which needs community participation.

Recommendations

It is a reality in Uganda that disasters due to severe climate events are on the increase, with a sizeable number of lives affected through deaths and loss of property. It is still a daunting challenge to mitigate such disasters more especially mudslides and landslides in Mt. Elgon areas in eastern Uganda. However, several measures should be undertaken:

1. The capacity for disaster mitigation should be enhanced through;
   a. Strengthening human resources (scientists and technical staff)
   b. Promoting science, technology and innovation
   c. Encouraging collaboration with neighboring countries and research centers
   d. Increasing resources in terms of finance for disaster risk mitigation
   e. Organizing national and regional for a concerted approach to disaster mitigation
2. The database for disaster management and mitigation should be organized through;
   a. Identification of prone areas for landslides and mudslides
   b. Periodically update the relevant data on trends of affected areas.
   c. Aiding demarcation based on proven data
   d. Sharing of climate change knowledge with different stakeholders
   e. Facilitating professionals to generate disaster-specific data for purposes of targeting severe risks
3. Demarcation should be streamlined through:
   a. Implementing policy measures that enhance technical guides for demarcation

b. Strengthening institutions that coordinate and implement demarcation activities

c. Enforcing laws and regulations related to demarcation

d. Establishing a sustainable mechanism for identifying and availing funds for demarcation activities

e. Ensuring that local communities are involved in demarcation exercises

4. Mobilization should be strengthened to;

a. Sensitize community to appreciate the value of leaving the affected areas

b. To drive the masses away from risk prone areas

c. To ensure appropriate land use practices in affected areas

d. To abide by the regulations governing the reserved areas affected with landslides and mudslides

e. Devise community mobilization methods to address the different needs of affected areas

5. Enforcement of laws and regulations should be enhanced through;

a. Ensuring coordination of stakeholders at ministry, district and lower level governments

b. Generation of and access to accurate data on risk areas and activities

c. Development of a checklist for tracking progress on enforcement activities

d. Prioritization of disaster preparedness enforcement into mainstream planning

e. Evaluation of the effectiveness of enforcement strategies in relation to risk reduction.

Abbreviations

DFID: Department for International Development

DISO: District Internal Security Officer

DPC: District Police Commander

DMC: Deputy Metropolitan commander

FAO: Food and Agricultural Organization

GHG: Greenhouses Gases

IPCC: Intergovernmental Panel on Climate Change

LC: Local Council

MPC: Metropolitan Police Commander

NAMA: Nationally Appropriate Mitigation Actions

NECOC: National Emergency Coordination and Operations Center

NEMA: National Environment Management Agency

RDC: Resident District Commissioner

RPC: Regional Police Commander

UNDP: United Nations Development Program

UNFCC: United Nations Framework Convention on Climate Change

UPDF: Uganda People's Defense Forces

References

1. Baastel (2014) Economic Assessment of the Impacts of Climate Change in Uganda- Draft Energy Sector Report. Gatineau, Canada; Baastel.

2. FAO, 2011. World census of Agriculture: analysis and international comparison of the results (2005-2011). FAO Statistical Development Series No. 13.

3. Fussel, K. and Klein, D. (2008) patterns of growth and public spending in Uganda: Alternative scenarios for 2003-2020. World Bank, Washington.

4. Kigenyi, 2014. Climate Issues in Changing Uganda. Fountain Publishers, Kampala Uganda.

5. Metroeconomica, *et al.*, 2015. Climate and Policy processes in Africa. AFRI, Abidjan, Ghana.

6. NEMA, 2008. State of the Environment Report for Uganda. Kampala, Uganda.

7. NAMA, 2015. The process in Uganda. Kampala, Uganda.

8. National Development Plan Uganda 2015/16 – 2020/21, National Planning Authority. Kampala, UgandaNational Planning Authority, Uganda Vision 2040. Kampala, Uganda.

9. The National Policy for Disaster Preparedness and Management (2010). Kampala, Uganda.

10. Uganda Bureau of Statistics (2013). Annual report. Kampala, Uganda.

11. UNDP (2015). Development Review on Climate Change. Washington, DC.

Technology for Disaster Risk Reduction

Chapter 14

Identification of Indonesian Technology Readiness in Disaster Risk Reduction

Adawiah[1], Ophirtus Sumule[2] and Irsan Pawennei[3]

[1]*Deputy Director,*
Harmonization of Innovation Programmes and Policies,
Ministry of Research, Technology and Higher Education,
Jakarta, Indonesia
E-mail: adawiah@ristekdikti.go.id, adawiah67@yahoo.com
[2]*Ministry of Research, Technology, and Higher Education,*
Republic of Indonesia
[3]*Centre for Innovation Policy and Governance*
Republic of Indonesia

Background

Based on the factors of geography, geology, climatology, and demography, Indonesia is prone to disaster risks. Badan Nasional Penanggulangan Bencana (BNPB) data recorded 1582 disasters have occurred throughout 2015 that resulted in 240 deaths, 1.18 million people were displaced, 24 365 homes were damaged, as well as 484 units of public facilities damaged. Landslide becomes the deadliest natural disaster throughout 2015 with 147 victims died. From the economic aspect, forest and land fires caused losses of up to Rp 221 trillion. Although the number of casualties and damage to buildings decreased compared to previous years, economic losses due to disasters in 2015 is the highest in Indonesian history.

In an effort to achieve the goals of Sustainable Development Goals (SDGs), a country's ability to prevent and cope with disasters is one key to the success of sustainable development and the overall economy. In a similar framework, Nawa

Cita (President Joko Widodo's Nine *Priorities* Agenda) of the 2015-2019 National Medium]Term Development also stressed the importance of enhancing the capacity to reduce disaster risk index, especially in the area of ??growth centers. So, while putting attention to the Framework for Disaster Reduction Sendai (Sendai Framework for Disaster Risk Reduction/SFDRR), integrated efforts and unified approach in various sectors of development for disaster risk reduction should be appointed into ashared spirit among stakeholders.

Government has a vital role as a driving force in improving community capacity while reducing bottlenecks in the utilization, the development, and the mastery of science and technology. To achieve this objective, the identification of needs should be undertaken, as well as the assessment of mastery, the transfer preparation, and the dissemination of disaster-related science and technology. In this context, cooperation between providers and users of disaster-relatedtechnology then plays a central role, given the technology must be adapted to the real needs on the ground.

BNPB identified 18 types of disasters in Indonesia, namely: earthquakes, volcanic eruptions, tsunamis, landslides, floods, flash floods, drought, fires, forest and land fires, tornados, tidal waves or storm surge, erosion, transport accidents, industrial accidents, extraordinary events, social conflict or social unrest or riot, acts of terror and sabotage. In the collection of statistical data and information on disasters in Indonesia, BNPB add two new disaster categories, namely: "earthquake and tsunami" for the category of the tsunami caused by the earthquake and "floods and landslides" to landslides caused by flooding. Thus, there are a total of 20 types of disasters in Indonesia.

Judging from the geographical conditions, the Indonesian archipelago is located at the confluence of three major tectonic plates, the Indo-Australian, Eurasian, and Pacific. In addition, Indonesia also has 142 volcanoes and coastline as long as 95180.8 km. Thus, according to this data, Indonesia has the potential to earthquakes, volcanic eruptions, and tsunamis. The entire island of Java, where about 56.8 per cent of Indonesia's population lives, has a high level of vulnerability.

In terms of disaster risk, Indonesia faces extensive risk disaster greater than the intensive risk type. Extensive risk disaster is associated with high frequency and predictable events, but only occur in a particular area, such as floods, landslides, or land and forest fires. This extensive risk usually occurs due to non-structural causes such as urban development that is not planned and managed properly, environmental degradation, poverty and economic disparity, vulnerable rural life and weak governance.

In facts, Indonesia has developed many early warning technology aimed at overcoming intensive risk disaster. However, the technology that can address extensive disaster is unfortunately underdeveloped.

Disaster management cycle describes a process that is continuously carried out by the government, the private sector and civil society to reduce the risk of disaster, to act quickly when a disaster occurs, and the process of recovery after a disaster. Appropriate action at all phases will lead to the increasing preparedness and better early warning, as well as to reduce the vulnerability of communities and to improve the disaster prevention when the cycle repeated. A complete cycle includes the formulation of plans and public policies to reduce the causes of disasters and reduce the impact on people, property, and infrastructure. This cycle consists of several phases, as follows:

1. Mitigation - reduce the effects of disasters. Example: developing building code and disaster zones, public vulnerability analysis, and public education.
2. Preparedness - to plan how to cope with the disaster. Example: drawing up preparedness plans, emergency training, early warning systems.
3. Response - efforts to reduce the threat created by the disaster. Example: search and rescue, emergency assistance.
4. Recovery - return the community to normal conditions. Examples: temporary shelter, cash assistance, medical assistance.

Cycle of disaster management has become a global benchmark in the field of disaster, including for Indonesia. Indonesian Disaster Management Bill suggests a paradigm shift in disaster management in Indonesia, of which initially focused on emergency response to disaster risk reduction. The consequence of this is a paradigm shift in disaster management as a series of measures that include the establishment of development policies that consider the aspects of disaster risk, disaster prevention, emergency response and post-disaster rehabilitation.

Related to the principles of science and technology in the response to this disaster, the Government Regulation Number 21 of 2008 confirms that science is needed in every phase of the disaster cycle (Figure 14.1).

Purpose and Objectives

This study intends to identify efforts to sustain the embodiment of disaster-related technologies independence and provide policy recommendations related to the utilization of innovation product in disaster risk reduction. The purpose of this study is as follows:

1. Mapping the current condition (existing condition) of disaster-related technologies that have been used and needed in Indonesia
2. Determining the focus of the technology needed in order to overcome the disasters, according to the level of disaster in Indonesia
3. Analyzing the profile of disaster-related technology to obtain sources of innovation product strategy of disaster, as well as its utilization.
4. Performing an analysis of the current condition to provide policy recommendations to support the development and utilisation of disaster-related technologyy.

Research Questions and Methodology

The research question in this study are as follows:

1. What is the existing condition of disaster-related technology in Indonesia?

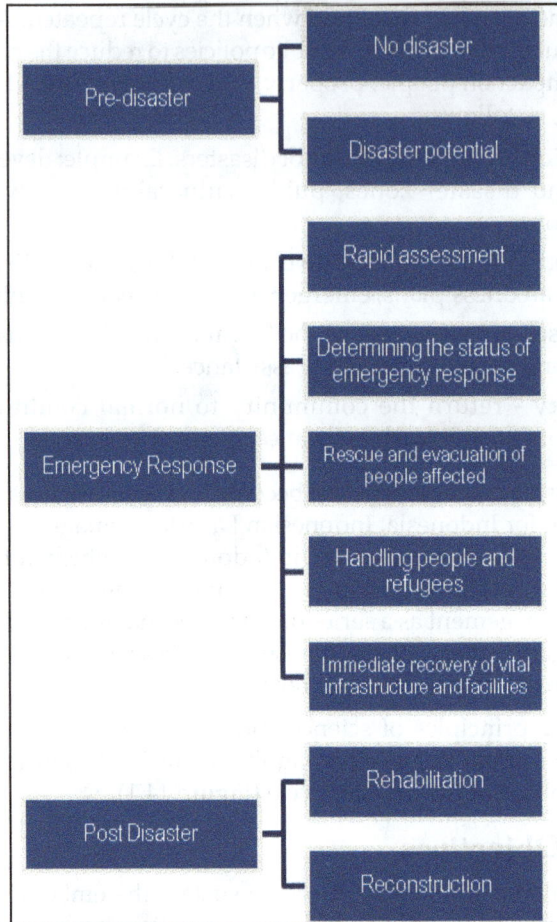

**Figure 14.1: Activities in Disaster Phases that
Need Science and Technology Support.**

2. What is the focus of the technology that needed in order to overcome the disaster, according to the level of the disaster in Indonesia?

3. What are the profiles of disaster-related technology focuses, including the sources of innovation and the users of the products?

4. Are there certain factors that affect the strategy of utilization of the technology? To what extent the existing policies support the development and the use of disaster-related technology?

The elaboration and methodology of above questions are summarized in the Table 14.1.

Outcomes of this study requires explorative-descriptive research. The exploratory nature of the research requires a comprehensive search in order to paint a complete picture of the object of the study. While the descriptive nature of the

Table 14.1: Matrix Research Questions and Research Area

No.	Outcome	Research Question	Research Area	Methodology
1	Disaster focus	What types of disasters are the focus of disaster-related studies in Indonesia?	Statistics on the number of events and the impact of disasters in Indonesia	Literature study
2	Disaster-related technology in Indonesia	What technology is needed when disaster strikes?	Mapping the technology that has been generated by research and development institutions and/or higher education.	Literature study Interview Focus group discussion (FGD) Disaster Risk Reduction Week
3	Focus of disaster-related technology	What technologies should be prioritized for development by technology providers?	Looking for a list of technologies that have been developed, its technology providers, and the level of preparedness of each technology with a scale of TRL (technology readiness level).	FGD Interview Literature study Tabulation of technology Group Interview
		Who is able to provide the technology?		
		How is the technology readiness that has been and is being developed by disaster-related technology provider in Indonesia?		
4	Policies that support the use of technology	How disaster-related technology development strategy that is appropriate to be applied in Indonesia?	☆ In general, the strategy of development and diffusion of each technology is formulated, as well as the stakeholders involved.	FGD Interview
		How each stakeholder can play a role in disaster-related technology diffusion strategy?	☆ Identifying related policies that support the development and difusion of technology	
		What are the policies that need to be made in this framework?		

discussion showed the object of study in detail and depth. Two of these properties resulted in this study used a qualitative research approach.

Qualitative research aims to study a phenomenon as a whole, depth, and detail. This is achieved through the excavation of qualitative data in the form of perspectives, opinions, and secondary data from the available publications. Therefore, this study uses data collection instruments such as in-depth interviews, focus group discussions, group interviews, and review of the literature.

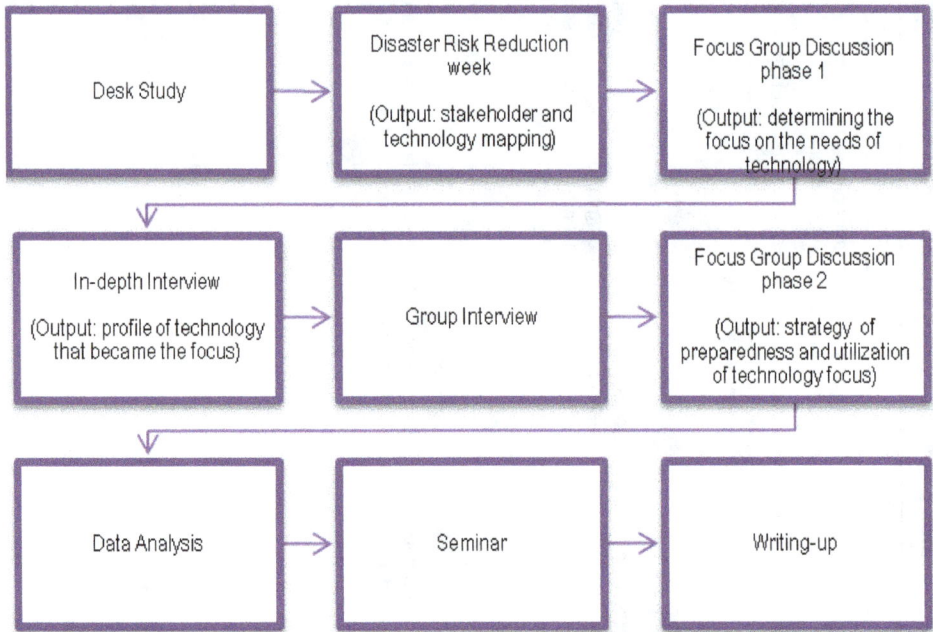

Figure 14.2: Flowchart of Research Phases

The period of data collection was done from September to early December 2015.Tabulation of the technology is updated every time there is a new input as the results of focus group discussions, interviews, or a desk study. Overall, there are four times the tabulation technology were updated to the final version of which is detailed in this report. The complete document concerning the history of technology tabulation at each phase is kept by researcher and can be demonstrated if needed.

This tabulation is based on disaster focus that has been chosen as the scope of this study. There are four disasters focused in this study that obtained from the mapping result on the data of casualties and the frequency of occurrence of the disaster, namely the earthquake and tsunami (this group include: earthquakes, tsunamis, and earthquakes and tsunamis); volcanic eruptions; floods and landslides (this group include: flooding, landslides, and floods and landslides); as well as land and forest fires. The data used as the basis of this choice is the data from the 1815 to 2015 time span.

Technology Readiness Levels

TRL 0: Idea. Unproven concept, no testing has been performed.

TRL 1: Basic research. Principles postulated and observed but no experimental proof available.

TRL 2: Technology formulation. Concept and application have been formulated.

TRL 3: Applied research. First laboratory tests completed; proof of concept.

TRL 4: Small scale prototype built in a laboratory environment ("ugly" prototype).

TRL 5: Large scale prototype tested in intended environment.

TRL 6: Prototype system tested in intended environment close to expected performance.

TRL 7: Demonstration system operating in operational environment at pre-commercial scale.

TRL 8: First of a kind commercial system. Manufacturing issues solved.

TRL 9: Full commercial application, technology available for consumers.

Figure 14.3: Technology Readiness Levels.

Source: European Commission (http://serkanbolat.com/2014/11/03/technology-readiness-level-trl-math-for-innovative-smes/).

Classification of technology product is adopting the concept of Technology Readiness Level(TRL). TRL helps indicate the status of the technology being developed. TRL consists of a 1-9 scale, with TRL 9 understood as a technology that has been fully prepared to be commercialized. In this study, the range of TRL is used as a means to estimate the range of technological readiness and to identify gaps (if any) between the need and availability of technology in disaster issues. Determination of the range of technologies is set based on the literature and based on the views of the relevant technology providers. The views and clarification on TRL range were conveyed when performing a series of FGD with stakeholders in disaster-related issues.

Results and Discussion

Of the 50 technology products (Figure 12.4), technology that is most widely mapped is technology for the prevention and preparedness as many as 37 products. Disaster risk management strategy chosen by most of stakeholders is a strategy that can reduce the vulnerability of the population to the risk of disaster, one of them is through the implementation of early warning systems.

Taking into account the above technology products (Figure 12.4), the stakeholders selected 12 technology focuses. Stakeholders assess these technologies to be continued and developed, among others due to, but not limited to:

1. Some of the technology has been utilized and responded positively by users, in this case are BNPB and other stakeholders. The closeness between providers and users will also be aspecial consideration.

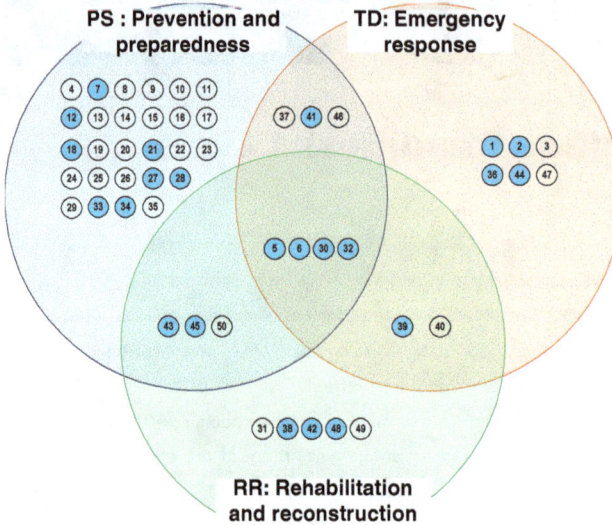

Products are expected to bein the range of 8-9TGL. Determining the range of technology based on the results of the literature study and clarification of the stake holders in a series of FGD.

Figure 14.4: Availability of Disaster-related Technology in Indonesia.

1. Sea water desalination (reverse osmosis); 2. Water Treatment Quick and Independent; 3. Processing of potable water (consisting of raw water pumps, pressure filters, filter manganese zeolites, activated carbon filters, cartridge filters and ultraviolet sterilizer); 4. Sadewa - Satellite Disaster Early Warning System; 5. Drone (Puna Wulung); 6. Drone Rajawali 330 (military specification-Defense); 7. Community preparedness – sirens; 8. The temperature sensor; 9. The Jampang (Information Systems and Puddle Rain); 10. Flowmeter; 11. FEWEAS (Flood Early Warning Early Action System); 12. Tech4Water (Flood Early Warning System); 13. EWS flood simple community, using accumulator, TOA, and sirens; 14. The flood EWS mobile apps; 15. Network accelerometer; 16. Digital Seismograph Short Period; 17. FDRS (Fire Danger Rating System); 18. Application Sipongi (system of early detection of forest fires); 19. FRS (Fire Risk System); 20. Detection of landslides with inclinometer and accelerometer; 21. Landslide EWS with extensometer, tiltmeter, a graduated rain (rain gauge), sirens and repeaters (EWS Gama, Sipendil); 22. Geulis (Geo-scince Landslide Early Warning and Information System); 23. Landslide EWS with extensometer, tiltmeter or accelerometer, rain gauge (a graduated rainfall), soil moisture, sensors and sirens parent; 24. EWS landslides community with wire extensometer; 25. Landslide EWS with extensometer, tiltmeter, and rain gauge; 26. Ina-Donet (Dense Ocean-Floor Network System for Earthquake and Tsunami); 27. Ina-TEWS (Indonesia Tsunami Early Warning System) with Buoys, sea level monitoring (satellite altimetry), tide gauge/mareograph/marigraph; 28. Lidar (Light Detection and Ranging); 29. Marine Radar (Radio Detection and Ranging); 30. RISHA - Simple Instant Healthy House; 31. Light and earthquake resistant house; 32. SIMBA - Information System for Disaster Mitigation; 33. SRGI (Indonesian Geospatial Reference System); 34. WRS - Warning Receiver System; 35. Saves Artificial Rain Water Aquifers; 36. Bisku Neo; 37. RDS FM (Frequency Modulation - Radio Data System) FM Radio; 38. Revetmen; 39. Rural BTS; 40. PLTS Hybrid (solar and diesel); 41. Weather Modification Technology (Casa aircraft, radar); 42. Doors automatic water corrosion resistance; 43. Erosion protection on roads (concrete mat, vetiver grass, bronjongan); 44. Communication equipment HF/UHF; 45. Lava channel dodger; 46. Emergency call out/push button emergency mobile apps; 47. Water bombing - water pressure at a depth of peat; 48. Biopori for agriculture after the eruption (sandy soil); 49. Innovation in the construction of houses on the beach; 50. Technology peat fertilizer (*e.g.* Bio-charging, microbial consortia, pugas).

2. Although some have been applied in the field, some arethe products that not yet fully functioning optimally. They need to be developed in order to further reduce the risk of disaster.

3. The technology should accomodate the values of local wisdom.

Although some of these technologies has reached a high level of readiness (TRL 8-9) and has been applied in disaster management, but others still an initial idea and have not been proved in the actual field. One example of this type of product is the mild earthquake-resistant homes that are being developed by the University of Syiah Kuala in Aceh.The list of all 12 technologies that become the focus of this study, including their position in the four disaster focuses is presented in Table 2.The brief profiles of the 12 technology focuses can be described as follows:

1. **Water Treatment Quick and Independent (Pusair PUPR):** It is a product for the purification of water that can be used emergency response. This self-contained water treatmentis available in sachet packagingthat so practical touse.

2. **Sadewa–Satellite Disaster Early Warning System (LAPAN):** An early warning system to predict the occurrence of extreme weather. The platform is able to provide close to real time information about potential extreme rainthat can lead to floods and landslides to a radius of 5km. Sadewa developed based on the observation satellite MTSAT (Multifunctional Transport Satellites). Sadewa will continue to be developed in order to be able to produce predictive information from one to three days ahead with the resolution remains high and is report every hour. Sadewa1.0 was developed in 2010 as a pilot project for the area of West Java, and later developed Sahadev2.0 for Indonesia. Currently, this platform has reached Sadewa3.0.

3. **Landslide Early Warning System (UGM):** There are two early warning system developed by the University of Gadjah Mada (UGM), the GAMA-EWS and Sipendil. GAMA-EWS tool patents have been registered since created in 2007. As of 2013, this tool has been installed in more than 12 provinces in Indonesia, in collaboration with the BNPB, ICL-UNESCO, Ministry of Rural Development (the naming of the time), and a number of NGOs and companies mining and oil and gas. This tool detects the rift within the soil, surface deformation and intensity of rainfall to determine the potential for the occurrence of landslides. In 2015, Gama Multi Group is believed to produce this tool.

4. **The Sipendil:** A landslide early warning system that consists of four major components, namely: the collector, the siren, the reservoir tube, and a controller box. Sipendil is based on rainfall and threshold can be set at some threshold according to the character of each region. Sipendil is reasonably priced (sold at Rp 1,000,000, -) and can be made by the community. Sipendil is currently in the process of patent registration.

5. **Geulis - Geo-science Landslide Early Warning and Information System:** An early warning system is the result of collaboration between

LIPI and Japan Radio Co. Ltd. (JRC), which features a combination of information intensity of rainfall, slope hydrological conditions, as well as the movement of the ground to generate a warning to the community.

6. **Digital Seismograph Short Period (BMKG):** As the name implies, this tool serves as a registrar magnitude earthquake when the earthquake strikes. This tool has been installed in areas prone to earthquake and tsunami as part of disaster response efforts. Digital Seismograph Short Period consists of Seismic Data Digitizer and Recorder, Short Period seismometer type triaxial 3 components (velocity sensitive sensor) type TDS-303 (3 parts) with a sampling frequency of 100Hz, GPS Timing, solar panel, and cabinet accessories with designs robust and water proof to store two units of solar charging panel, and antenna rod GPS antenna, cable and compass. Also equipped with software processing and analysis of earthquakes using Windows OS, automatic detection and alarm if there is an earthquake.It can process and analyze the location of an earthquake wave spectrum of seismic signals.

7. **In a TEWS –Indonesian Tsunami Early Warning System (Research coordinator):** It is a national project involving various institutions in the country under the coordination of Research and Technology as the Tsunami disaster early warning efforts. In a TEWS is an advanced seamonitoring system that was operated. However, currently it is invery poor condition. Of the 25 buoys attached, are currently only staying 3 buoys whose condition was questionable. In fact, the benefits In a TEWS will not only be felt by the people of Indonesia, but the international community, especially those located in the Indian Ocean and the South West Pacific and South China Sea.

8. **RISHA –Simple Instant Healthy House (Pusperkim PUPR):** RISHA is a livable and affordable housethat can be built to survive with knock down system based on modules. RISHA can be built in a relatively fast and with limited manpower. Due to the size of RISHA components refers to the modular size, so components of RISHA are flexible and efficient in their consumption of building materials. So far, RISHA has had 67 applicators and applied by more than 10,000 units of the post-Tsunami Aceh. The application of the model have been conducted in West Java, Riau Islands, West Sumatra, Bengkulu, Lampung, Bangka Belitung, Riau, Jambi, East Kalimantan, South Kalimantan, North Sulawesi, Central Sulawesi, West Sulawesi, East Java, Yogyakarta, Central Java, Bali, NTT, NTB.

9. **Light and earthquake resistant house (Syiah Kuala University in Aceh):** The house is liveable and affordable, with earthquake-resistant construction. Currently being developed by the University of Syiah Kuala in Aceh as a modification of RISHA. By several parties, the development process of RISHA is still considered toolong and needs to be improved.

10. **SIMBA - Information System for Disaster Mitigation (LAPAN):** A service of natural disaster mitigation information system based on remote sensing. This service is intended for stakeholders who need information

Table 14.2: The Readiness of 12 Disaster-Related Technology Focuses According to Stakeholders

No.	Technology Products	TRL Prediction	Earthquake and Tsunami			Volcanic Erruption			Flood and Landslide			Forest and Land Fire		
			PS	TD	RR	PS	TD	RR	PS	TD	RR	PS	TD	RR
1	Water Treatment Quick and Independent (Pusair PUPR)	8-9		x			x	x		x				x	
2	Sadewa/LAPAN	6-7	x			x			x			x		
3	Landslide EWS/UGM	8-9							x					
4	GEULIS/LIPI and JRC	6-7							x					
5	Digital seismograph short period/ BMKG	6-7	x			x			x					
6	InaTEWS/Ristek (koordinator)	8-9	x											
7	RISHA/Pusperkim PUPR	8-9	x	x	x	x		x	x		x			
8	Light and Earthquake Resistant House/Uni. Syiah Kuala Aceh	3-5			x		x	x		x				
9	SIMBA/LAPAN	8-9	X	x		x	x		x	x		x	x	
10	Weather Modification Technology/BPPT	8-9								x			x	
11	Lava channel dodger/PUPR	8-9				x		x				x		
12	Peatland fertiliser/BPPT	3-5												x

PS: Prevention and preparedness; TD: Emergency response; RR: Rehabilitation and reconstruction.

about the condition before, during and after the disaster. The main data used is data satellite Terra/Aqua MODIS, NOAA AVHRR, MTSAT-1R, QMorph, and TRMM.

11. **TMC (Weather Modification Technology):** It is an effort of human intervention in the control of water resources in the atmosphere, which aims to minimize natural disasters caused by climate and weather through the utilization of weather parameters. TMC can be used to increase precipitation (used to fill the reservoirs and drought) or reduce the intensity of the rainfall (as flood prevention) in certain areas. The intervention is done through the air doing the stocking material at the nursery at a certain altitude clouds.

12. **Lava Channel Dodgers (PUPR):** As the name implies, this product is part of the handling of the removal of the post-cold lava eruption of the volcano. The existence of this channel allows overflow of material from the upstream area to be contained so there is no overflow into the surrounding area.

13. **Peatland fertiliser (BPPT):** Peatland fertiliser is a technology to convert peatland that have been damagedso that fertile again. The existence of this technology allows the dry land and marginal by improving soil structure and increase the availability of minerals that can play a role in food sufficiency.

All technology focuses are of course a form of proposals from stakeholders involved in this study. In the end, the final decision is in the hands of decision-makers to determine which technology will be the focus of development. Human is the major component in disaster issues (Figure 14.5 and Table 14.3). Disaster-related technology is not the technology developed for the purpose of fulfilling the curiosity of researchers or technology providers only. Disaster-related technology is developed to increase the capacity and the human ability to cope with and reduce the risk of disaster.Its users not only consist of a single party, but multi-users. BNPB is the most powerful in the constellation of disaster-related technology users, but keep in mind, there are others who also need to harness the technology in disaster. Some of them include: the community (who are vulnerable and living in disaster prone areas), civil society organizations, and local government. Although other users are not the biggest investor in technology development, but the role and importance are also considerable.

Meanwhile, technology diffusion strategies of disaster can be shared by stakeholders, namely: government, technology providers, and users of technology, as in the Figure 14.5.

Conclusion and Recommendation

According to the mapping of the disaster related technology in Indonesia, most of the available products are technology for prevention and preparedness stage that predicted having TRL 8-9. This is in line with the paradigm of disaster risk reduction in SFDRR, that is also echoed by the Government's Bill in Disaster Management, in

Table 14.3: Development Strategy for Disaster-Related Technology in each Range of TRL Prediction

TRL Prediction	Phase	Development Strategy of Disaster-Related
0-2	Basic Research	1. Human a main component. 2. User engagement and assessment. 3. Collaboration withv arious parties with a variety of disciplines. 4. Understand the vulnerability of communities to disasters (intensive and extensive risk). 5. Ensure timely and accurate data issued. 6. Develop a target and a clear indicator of technology development.
3-5	Laboratory and field tests	1. Manas a main component. 2. User engagement and assessment. 3. Collaboration with various parties with a variety of disciplines. 4. Understand the social, economic, political, and culture of the target community of technology users. 5. Utilization and updating dataprecicely and accurately. 6. Monitoring and evaluation of each step of technology development.
6-7	Prototype system and Demo	1. Manas a main component. 2. Multi-stakeholder. 3. Collaboration with various parties with different disciplines 4. Lessons learned are well documented. 5. Utilization and updating data with precisely and accurately. 6. Monitoring and evaluation of each step of technology development.
8-9	Proven and commercialized	1. Manas a main component. 2. Multi-stakeholder. Especially, the need to involve the industry. 3. Collaboration with various parties with different disciplines. 4. Understand the social, economic, political, and culture of the target community of technology users. 5. Ensure training and public education on the technology that is being/has been developed. 6. Utilization and updating data with precisely and accurately. 7. Lessons learned are well documented 8. Monitoring and evaluation of each step of technology development.

which efforts should be put more into prevention and preparedness. Support for increasing the diffusion and adoption of disaster-related technology in the country can be done with policies which address the following issues:

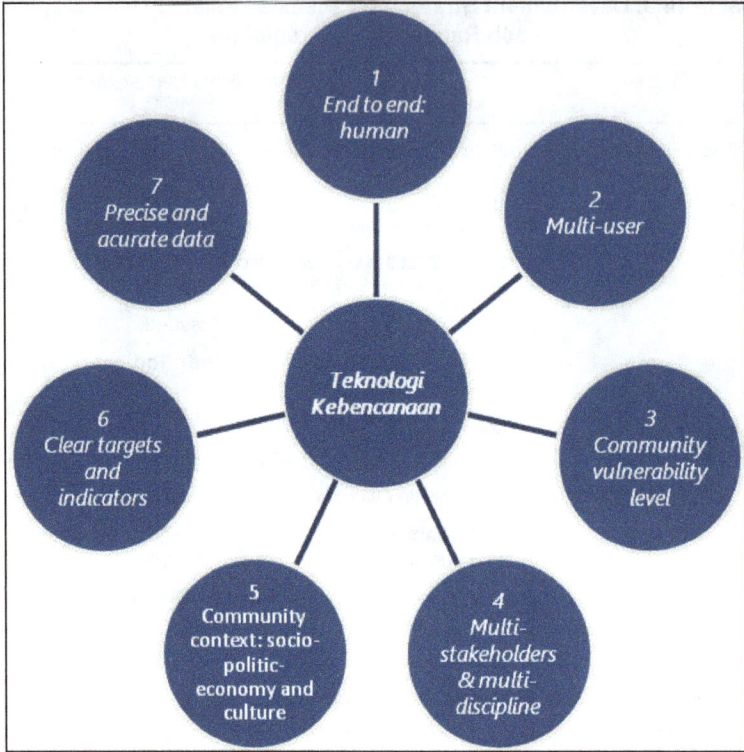

Figure 14.5: Components of Disaster-Related Technology.

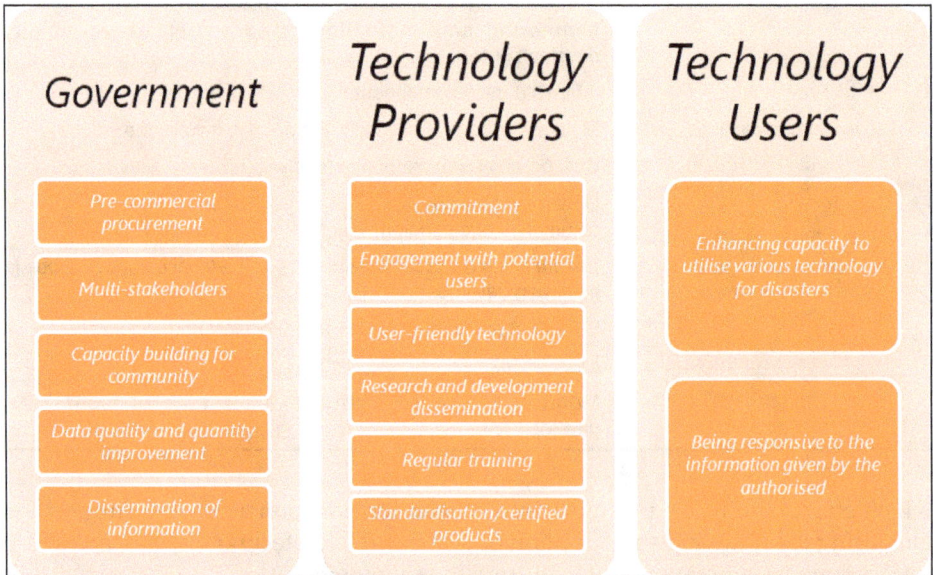

Figure 14.6: Technology Difussion Strategies for Stakeholders.

☆ Mainstreaming disaster issues in research and development, by attaching the issue of disaster into a National Research Agenda. This needs to be accompanied by an increase in research and development incentives in the field of disaster.

☆ In conducting the research and development of technology for disaster issues, it requires user assessment, including and especially in the social, economic, political, cultural of the community at risk.

☆ Social analysis is also needed to ensure that vulnerable communities are able and willing to adopt the technology, and attention should be put that each region has different characteristics and culture, where it takes an adjustment.

☆ Collaboration with other technology providers.

☆ Development of the technology should include the increased capacity and capabilities of potential users of the technology.

References

1. Government Regulation Number 21 Year 2008 about Implementation of Disaster Management.

2. Law Number 24 Year 2007 about Disaster Management.

3. UNISDR (United Nations Office for Disaster Risk Reduction), 2004. *Living with Risk: A global review of disaster reduction initiatives*. Geneva, Switzerland: UNISDR.

4. UNISDR, 2005. *Hyogo Framework for Action 2005–2015: Building the Resilience of Nations and Communities to Disasters*. World Conference on Disaster Reduction. Dapat diakses di: http://www.unisdr.org/2005/wcdr/intergover/official-doc/L-docs/Hyogo-declaration-english.pdf.

5. UNISDR. 2009a. *UNISDR Terminology on Disaster Risk Reduction*. Geneva, Switzerland: UNISDR.

6. UNISDR. 2009b.*Reducing Disaster Risk through Science: Issues and Actions*. The Full Report of the ISDR Scientificand Technical Committee 2009.

7. UNISDR. 2015. *Making Development Sustainable: The Future of Disaster Risk Management*. Global Assessment Report on Disaster Risk Reduction. Geneva, Switzerland: UNISDR.

Chapter 15

Confiability Analysis of the Automatic Weather Stations Network of Sri Lanka

Nuwan Kumarasinghe

Senior Electronics Engineer,
Department of Meteorology,
Colombo, Sri Lanka
E-mail: nuwan1960@gmail.com

ABSTRACT

Sri Lanka is exposed to several natural hazards such as flood, drought, heavy rainfall, wind storms, landslides and lightning which are related to hydrology, meteorology and prevailing weather conditions.

Heavy rain is very common and often results in flash floods, landslides and rock falls which are potentially very damaging for settlements and public infrastructure. Dissemination of rainfall and other meteorological information in timely and accurate manner is a challenging task for meteorologists. Many of the advances which have been made in disaster risk reduction rely heavily on modern technology. Information obtained from automatic weather stations (AWS) plays significant role in weather forecasting. With the advancement of modern technology, information gathered from AWS can be disseminated in more frequency. Among the other communication means, very small aperture terminal (VSAT) which is satellite based communication media useful specially for disastrous situation, where terrestrial based communication links are hardly available. The reliability of these information or data heavily depends on sensors, processing and site selection. Application of modern science and technology has a major role to play in ensuring that lessons learned are applied. Science and technology's continued advancements and research have allowed scientists to apply strategies and policies to mitigate risks and build resilience to natural and human made disasters.

This paper discusses, technological and unexpected environment challenges encountered when automatic weather station networksare established in a department of meteorology.

Keywords: *Automatic weather station (AWS), Very small aperture terminal (VSAT), Sensor and lessons learned.*

Introduction

Disasters often follow natural hazards. Severity of a disaster depends on how much impact a hazard has on society and the environment. The scale of the impact in turn depends on the choices we make for our lives and for our environment. In Sri Lanka, over 90 per cent of natural disasters are weather or climate related. Every year, floods, droughts, thunder and lightning activity and strong winds cause disasters in various parts of the island. The recurring damages due to climate and weather related disasters in Sri Lanka have increased during the recent decades, and the losses in terms of livelihood, food availability and economy for his kind of events are astoundingly high. Heavy rain is very common and often results in flash floods, landslides and rock falls which are potentially very damaging for settlements and public infrastructure. Dissemination of rainfall and other meteorological information in timely and accurate manner is a challenging task for meteorologists. Many of the advances which have been made in disaster risk reduction rely heavily on modern technology.

Automatic Weather Station (AWS)

Automatic Weather Station is the generic name for an automatic meteorological data acquisition system. Information obtained from AWS plays significant role in weather forecasting. AWS can issue weather information in very short time period (like in every 10 minute intervals) and this high frequent issuance of weather parameters is immensely useful for early warning purposes and this may reduce disaster risk. Basic block diagram of AWS is shown in Figure 15.1. Some AWS are installed for short-term projects (*e.g.* animal health emergency monitoring or near wild fires), some are installed for long-term projects (*e.g.* studying climate change). Some AWS are required to provide data in real-time (*e.g.* for irrigation), some provide delayed reports (*e.g.* for climate monitoring).

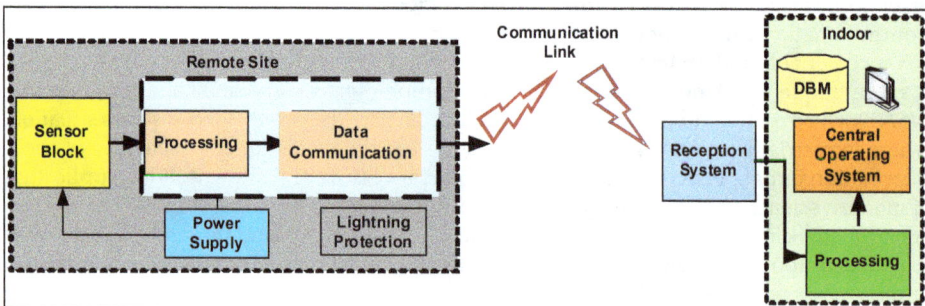

Figure 15.1: Basic Diagram of Automatic Weather Station.

Advantages of AWS

The following advantages can be noticed in AWS as compared to conventional measurements.

- ☆ Standardization of network observations, both in time and quality;
- ☆ Real-time continuous measuring of parameters on a 24/7 basis;
- ☆ Greater accuracy;
- ☆ Generally more reliable;
- ☆ Conducts automatic data archiving;
- ☆ Provides higher data resolution;
- ☆ Collection of data in a greater volume;
- ☆ Adjustable sampling interval for different parameters;
- ☆ Generally free of reading errors;
- ☆ Generally free from subjectivity;
- ☆ Automatic QC applied during collection and reporting stages;
- ☆ Automatic message generation and transmission;
- ☆ Monitoring of meteorological data;
- ☆ Access to archived data locally or remotely, and
- ☆ Data collection in harsh climates.

Disadvantages of AWS

The following disadvantages can be seen in AWS.

- ☆ Limited area representation, an area of about 3-5 km around the sensor site;
- ☆ It is not possible to observe all parameters automatically as are done through a manual approach toward taking an observation or an approach whereby the automatic observation is augmented by a human observer, for example cloud coverage and cloud types;
- ☆ More intense ongoing periodic routine maintenance;
- ☆ Increased periodic testing and calibration;
- ☆ Insure that a staff of well trained staff and specialists is maintained;
- ☆ Resulting higher cost of instrumentation and operation. However, efficiencies gained through greater levels of automation may result in some cost benefits.

Sensors

Sensor block is the heart of the entire system where different sensors are used to acquire meteorological parameters such as wind speed, wind direction, temperature, relative humidity, pressure, solar radiation and rainfall. When choosing a sensor, one has to consider environmental factors, economic factors and sensor characteristics.

The main environmental factors are to be considered are; temperature range, humidity effects, corrosion, over range protection, susceptibility to electro- magnetic interferences and ruggedness. Cost, availability and lifetime are the main economic factors to be considered. Sensitivity, range, stability, repeatability, linearity and response time are some common sensor characteristics. It is very important to consider all above mentioned factors during the design and installation stages. Unexpected results may obtain if improper selection is made.

Different Types of Sensors

There are three (3) types of sensors namely, analog, digital and intelligent type. Output of an analog sensor can be obtained in form of voltage, current, charge, resistance or capacitance. In a digital type sensor, the output will contain in a bit or group of bits, pulse or frequency. In intelligent sensor, a microprocessor provides an output in serial digital data or in parallel data form. A combination of all these three types of sensors can be seen in present day AWS.

Processing Unit or Block

Output of meteorological sensors used in AWS has different signal levels. The processing block amplifies these small signals in TO? meaningful level in order to ALLOW FURTHER process???. The main function of the processing unit consists of Data Acquisition, Data Processing,Data Storage and Telemetry. Processing unit takes regular data samples from different meteorological sensors. After taking samples, computation of averaging values over specified periods such as 1 minute, 10 minute, hourly or daily will be done. Take records of extreme values, example: maximum/minimum temperature or wind speed. After that, meteorological reports will be generated. Date and time stamp will be added for each data packet. Control signal will be generated to send back data through communication link. Temporary data storage will also be GUARANTIED in case OF A failure communication link. Backlog mechanism is available to resend data when communication link resumes. Automatic clock adjust mechanism to maintain synchronization of time with other remote AWS.

Timing Requirements for the AWS Processing Unit

The following timing is required for the processing unit.

Data sampling is normally done in every 1-sec, 4-sec. or 0.25 second period, depending on the sensor.Unit conversion is also take place for every sample. Periodic averaging will be done in 1-minute, 10-minute, hourly, and daily. Store minimum and maximum values, specially temperature, wind speed and wind direction (periodic). Report generation normally occurs in every 1-minute or 10-minute period. Data communication will take place in real-time and backlog data.Operator shall have the facility to interface at any instant.

Typical architecture of the processing unit is sown in Figure 15.2.

Processing block consists of multifunction microprocessor boards. Multiple analog to digital converter inputs for taking data sample from sensors. Signal processors are used for converting sensor input to analogue voltages. Special

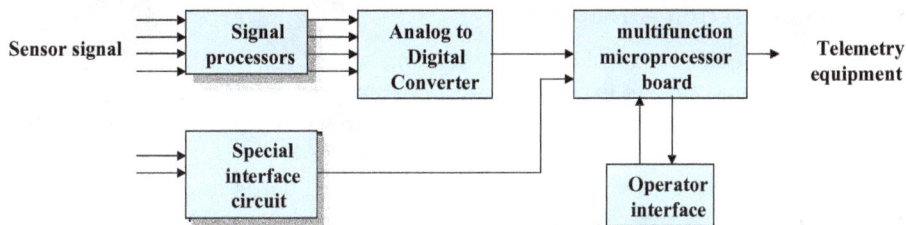

Figure 15.2: Typical Architecture of a Processing Block.

interface circuit is used for special sensor inputs like rain gauge. Operator interface is also available.

Communication Block

Meteorological data exchange between the remote AWS and the central location can be done using different communication means such as public telephone switch network, internet, mobile network, radio frequency, satellite and fiber optic. Communication block prepares the data packet which is ready to communicate between the remote station and the central location. AWS data telemetry network is shown in Figure 15.3.

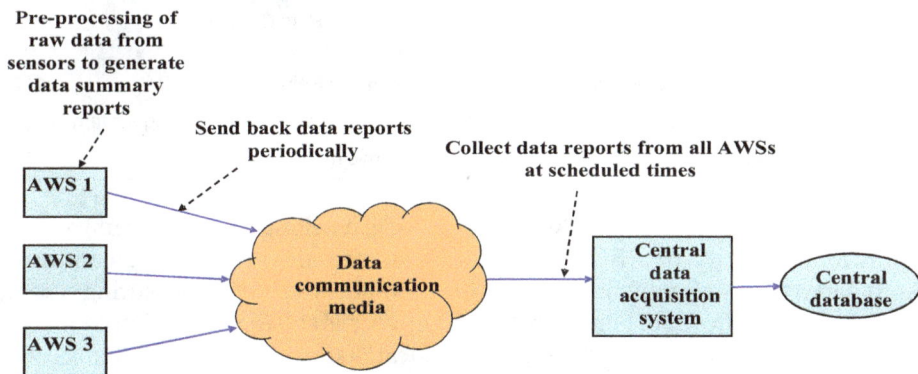

Figure 15.3: AWS Telemetary Network.

Present day AWS network widely use Ethernet based communication link such as Internet Protocol and Virtual Private Network (IP/VPN), Global System for Mobile Network (GSM) and satellite based Very Small Aperture Terminal (VSAT) for data exchange.

Power Supply Unit

AWS is normally powered by non commercial power using either solar or wind power. Solar power is widely used among AWS networks.

Lightning Protection Arrangement

AWS network are vulnerable for lightning strikes. Lightning is a natural phenomena which brings extremely high current in a very short time either as direct strike or induced current. Wind sensor of the AWS is normally fixed in 10 m high metallic mast and therefore it is more vulnerable for direct lightning strikes. Sophisticated electronics which are in all processor boards and communication equipment are more prone for lightning strikes. Lightning finial shall be installed together with proper earthing arrangement to minimize the effect of direct lightning strikes. Surge protection devices (SPD) are used to protect power and data lines.

Automatic Data Acquisition Systems Used in Sri Lanka towards Disaster Risk Reduction

There are number of automatic data acquisition systems in Sri Lanka towards disaster risk reduction belong to different organizations. Distribution of automatic data acquisition systems is given in Table 15.1.

Table 15.1: Automatic Data Acquisition System in Sri Lanka

Sl.No.	Organization	Description	Qty
1.	Department of Meteorology (DoM)	Automatic Weather Stations (AWS)	38
2.	Department of Meteorology (DoM)	Automatic Rain Gauges (ARG)	20
3.	National Buildings Research Organization (NBRO)	Automatic Rain Gauges	75
4.	Department of Irrigation (DoI)	Automatic Rain Gauges	122
5.	National Aquatic Research Agency (NARA)	Real-time sea level monitoring system	3
6.	Geological Survey and Mines Bureau (GSMB)	Seismic monitoring system	3

33 AWS out of 38 and 16 ARG out of 20, which belong to DoM are in working condition. Majority of ARG belong to NBRO are also in working condition. Automatic rain gauge network, some with automatic water level measuring sensors belong to DoI is shown in Figure 15.4. This network is still in verification condition. Network consists of; 122 hydro-meteorological stations throughout the country with automated data collection and transmission system anda central unit for data processing and analyzing at Colombo. This is a World Bank funded project. Out of 3 real time sea level sensors at NARA only one is in working condition. Seismic monitoring system of GSMB is presently in working condition.

AWS Network in Sri Lanka

Thirty eight (38) AWS were granted to Department of Meteorology (DoM), Sri Lanka under Japanese grant aid project (JICA) in 2009. Twenty two (22) AWS have been deployed in regional meteorological office premises including Colombo while the rest of AWS were installed in other stake holder institutions such as agriculture, plantation, universities or wild life premises. Six sensors, wind, temperature, relative humidity, solar radiation, pressure and rainfall are equipped for the AWS at meteorological premises. Four sensors, wind, temperature, relative humidity and rainfall are equipped for other AWS. Performance of the entire AWS network

Figure 15.4: Hydro Meteorological Information System.

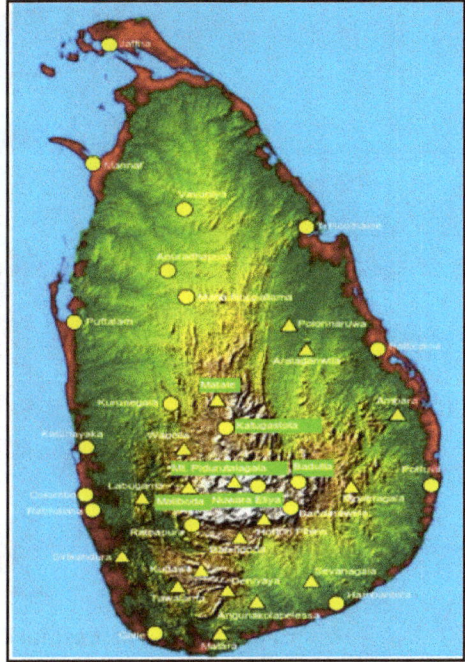

Figure 15.5: Automatic Weather Station Network.

including VSAT communication network was studied during the period of July 2009-December 2015. AWS network is shown in Figure 15.5.

Methodology

This study is mainly concentrated on the performance of AWS installed at 21 meteorological stations. The functionality of entire AWS network depends on reliability of individual components. Performance of sensors, data logger, VSAT communication equipment and earthing system was verified two times per year at every remote location. Data reception at central location in Colombo, data transmission at each remote station and reliability of VSAT communication link were tested.Data logger sends meteorological data packet in every 10 minutes to the central location at Colombo and these ten minute data collected at Colombo, was compared. There are over 315,360 data sets available for the period of July 2009 to July 2015. Once sensors, VSAT network, data logger, VSAT Hub and data base are in order, the entire network is in full operation. This is illustrated in Figure 15.6. Possible causes for each faulty unit were verified. Continuity of data from all sensors is concerned at all AWS locations. Health parameters of each AWS is also being sent together with meteorological data to the central database.

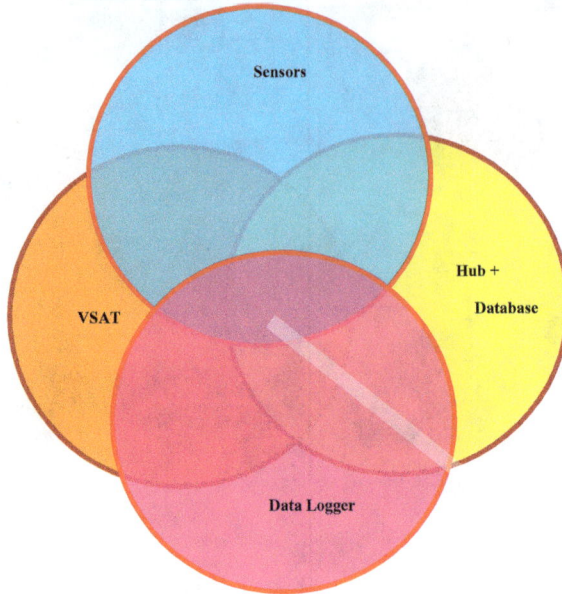

Figure 15.6: Successful Operaton of AWS Network.

Results and Discussion

Behavior of sensors during July 2009 – July 2015 is given in Figure 15.7. It is clearly noted that ultrasonic wind sensor is the most critical sensor in the system. Relative humidity sensor also has low performance as compared with other sensors.

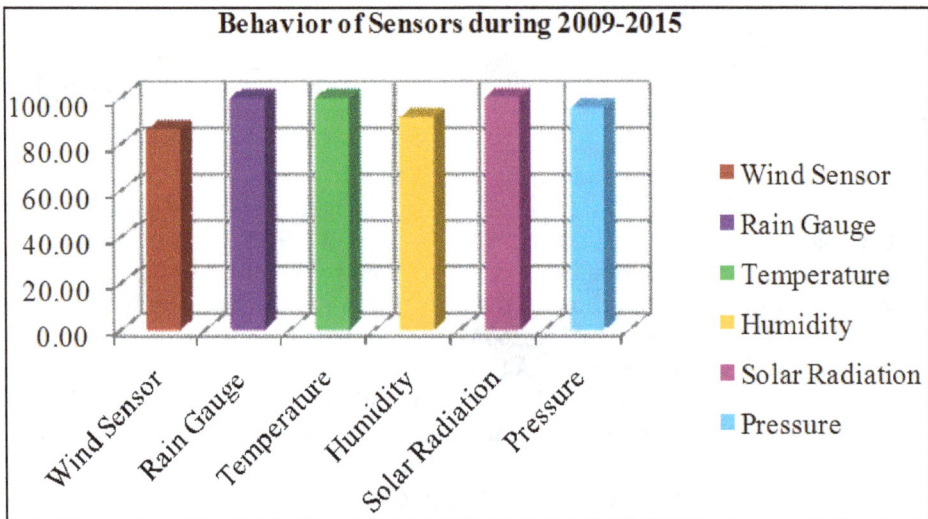

Figure 15.7: Behaviour of Sensors during 2009 July–2015 July.

It is noted that wind sensor which is ultrasonic type is the most critical sensor among others. Behavior of ultrasonic wind sensor at each AWS site is given in Figure 15.8.

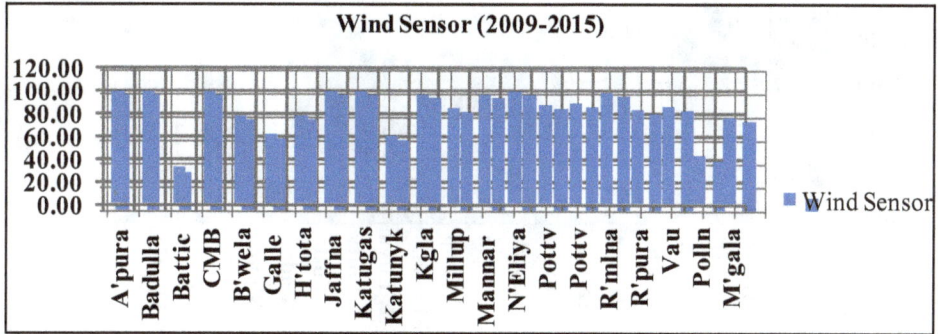

Figure 15.8: Performance of Wind Sensors at each AWS Site.

Wind sensors at Batticaloa, Galle, Katunayake and Pollonnaruwa are the most critical. Data logger performance at 21 AWS sites is shown in Figure 15.9. Data loggers at Galle, Mannar and Pollonnaruwa have comparatively low performance.

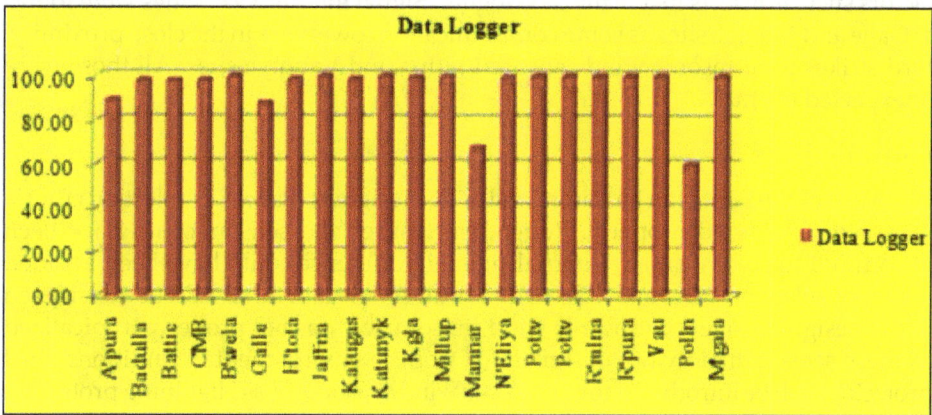

Figure 15.9: Performance of the Data Logger at each AWS Site.

Performance of VSAT at each AWS site is given in Figure 15.10.

VSAT at Galle, Jaffna and Pollonnaruwa has less than 50 per cent performance as compared with other sites.

The faults found can be categorized in the following manner.

i. Site specific   ii. Environmental

iii. Lightning   iv. External factors

v. Signal interference   vi. Unexpected

AWS at Galle and Mannarhave site specific issues such as trees and other objects obstruct the VSAT antenna paths. AWS at Galle and Hambantota which are

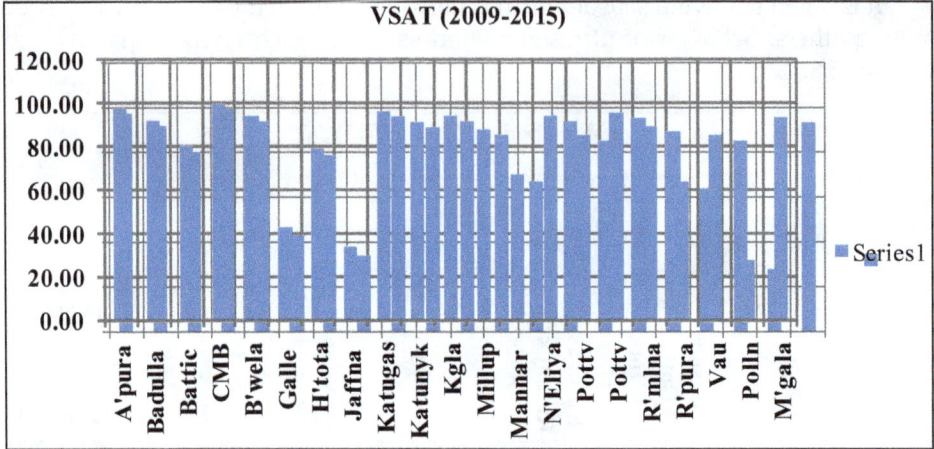

Figure 15.10: Performance of VSAT at each AWS Site.

close to coastal area,are prone to corrosion. Pollonnaruwa, and Rathnapura sites are more vulnerable for lightning. AWS at some sites were affected by external factors such as insects inside the data logger. Signal interference issues were found at Galle and Mannarsites as some communication towers are in the close proximity. Bird strikes to ultrasonic wind sensor at Jaffna and Vauniya sites. All those were unexpected events.

Conclusions

Most of the faults can be minimized at the designing stage. Site selection criteria were not considered properly. Rugged enclosures and sensors could have been introduced to coastal areas. External signal interference could have been avoided if proper selection was made. Security for all AWS stations should be considered in the designing stage. Ultrasonic wind sensor used was not properly tropicalized as these sensors mainly not designed for tropical environmental conditions. Bird protector shall be introduced for ultrasonic wind sensors. Class I lightning protection measures shall be introduced for lightning prone areas. Redundancy communication link shall be introduced for uninterruptible data acquisition. National guidelines for installation of automatic weather station or environmental data acquisition system shall be prepared specially for usage of DoM, National Buildings Research Organization (NBRO), Department of Irrigation and Department of Agriculture.

References

1. Meisei" AWS Operation and Maintenance Manual" Chapter 3, pp. 13,23,37.

2. WMO-No. 8 "Guide to Meteorological Instrumentsand Methods of Observation"- Part II, pp. 1-1-2-17.

3. Kumarasinghe, N., 2010. Coexistence of automatic weather station system in harsh environmental conditions at NMC Colombo, Sri Lanka. WMO-TECO, Helsinki, Finland.

Education and Research

Chapter 16

Academic Hubs: Using Applied Research and Community Services to Build Resilience of Nations and Communities to Seismic Disasters

Jalal Al-Dabbeek[1] , Hatim Alwahsh[2], Sami Sader[2],
Abdel Hakeem Juhari[2], Barbara Borzi[3], Fabio Germagnoli[3],
Paola Ceresa[4] and Ricardo Monteiro[4]

[1]*Urban Planning and Disaster Risk Reduction Center (UPDRRC),*
Najah National University (NNU), Palestine,
E-mail: seiscen@najah.edu
[2]*Najah National University (NNU), Palestine,*
E-mail: hatem_alwahsh@yahoo.com
[3]*European Centre for Training and Research in*
Earthquake Engineering (EUCENTRE), Italy
E-mail: barbara.borzi@eucentre.it
[4]*Institute for Advanced Study of Pavia (IUSS), Italy*
E-mail: paola.ceresa@eucentre.it

ABSTRACT

To reach the desired progress in science and technology, disaster risk reduction (DRR) Innovations, building resilience of nations and communities', the following key-elements are needed: data-collection, assessment, management methods, disaster risk institutional arrangement, risk communication, knowledge-sharing and group-work innovations.

Additionally, these elements enhance decision-makers managing DRs, implementing and monitoring particularly Sendai Framework for DRR 2015 - 2030.

Building resilience is a cross-cutting issue and a fundamental interdisciplinary concept, which requires natural and social sciences innovations. Building the resilience of nations and communities to disasters requires adopting holistic approaches concepts and tools.

An-Najah National University's Urban Planning and Disaster Risk Reduction Center (UPDRRC) brought players, using scientific knowledge and community services driving through change towards policy, preparedness and public awareness. UPDRRC, as an academic hub, has an important role in enhancing the Palestinian communities' disaster resilience, through adopting holistic approaches to DRR activities adopting scientific strategy drawing together decision-makers, practitioners and the public driving towards sustainable RR, exceeding traditional academic centers parameters. Based on networking and integration concepts and to achieve its goals and objectives, UPDRRC participated in and/or conducted several local, regional and international DRR projects and activities in cooperation with UN organizations and other scientific research institutes, such as: EUCENTRE and IUSS in Italy, GFZ in Germany, USGS in USA, *etc*. This paper contains examples for several dissemination activities as well as a detailed case study for an important scientific project "Support Action for Strengthening Palestine capabilities for seismic Risk Mitigation - SASPARM2".

Keywords: *Resilience, Academic hubs, SASPARM, Public awareness, DRR, Vulnerability, Innovations.*

Introduction

Resilience has a very important role in all disaster risk management phases and tasks: mitigation, preparedness, emergency response, recovery and reconstruction. So it's recognized as imperative for sustainable development and pivotal for implementing Sendai Framework for DRR 2015 - 2030 and priority for actions of UNISDR science and technology Road Map. Building resilience is a cross-cutting issue and a fundamental interdisciplinary concept, which requires natural and social sciences innovations. Building the resilience of nations and communities to disasters requires adopting holistic approaches concepts and tools.

The economic impacts of past disasters caused by natural hazards have been very great throughout history, and the potential future impacts may be even greater as populations grow and man's work expands. Economic losses from natural disasters have increased nine TIMES? over the last 50 years. During the 1960s, for example, economic losses from natural disasters were estimated at 73 billion US dollars; the decade of the 80s brought losses of $204 billion. From 1990 to 1999, the total economic loss from natural disasters exceeded $630 billion.

Today, more than half of the global population resides in urban areas. By 2025, roughly, two-thirds of the world's inhabitants and the vast majority of wealth will be concentrated in urban centers. Many of the world's mega-cities characterized as those with populations exceeding 10 million, are already situated in locations already prone to major earthquakes and severe droughts, and along flood-prone coastlines, where the impacts of more extreme climatic events and sea level rise pose a greater risk to disasters (Making Cities Resilient Report 2012, UNISDR).

Urbanization happening in relatively smaller cities is also a concern, particularly in regions where existing infrastructure and institutions are not equipped enough to cope with disasters. The vulnerability of this new generation of urbanites will become a defining theme of disaster risk in the coming decades. Against this backdrop, the report observes two diverging trends relevant to strengthening urban resilience (Making Cities Resilient Report 2012, UNISDR).

Over the last three decades, natural disasters, particularly earthquakes, have become increasingly destructive as they affect large concentrations of population and property. The cost of replacing and repairing earthquake-damaged buildings is a significant drain on the economies of earthquake-prone countries. Therefore, it is imperative to achieve methods of reducing earthquake damage to an economically supportable level.

The State of Palestine (Occupied Palestinian Territories OPT) and the neighboring countries are adjacent to the seismically active Dead Sea Transform (DST) and its associated geological faults and, consequently, are vulnerable to earthquakes. The documented history of the region dates back to biblical times and provides wealth of descriptions of destructive earthquakes. Apparently, all populated areas in the region have suffered great destruction and high death toll, as a result of earthquakes located along the DST and its associated fault systems. Seismologic data obtained instrumentally since the beginning of the century includes high-magnitude events (M > 6.0), some of which have caused significant damage and great financial losses. The occurrence of a strong earthquake becomes a major threat to the safety, social integrity, and economy for the Palestinian people. The best remedy to REDUCE?? earthquake loss is the proper planning and safer buildings; "Earthquakes do not kill people, buildings do!" In addition to awareness and preparedness, there is always an urgent need to implement the up-to-date knowledge and data in the form of building regulations and codes.

Background and Problem Statement

The Mediterranean region, because of its geological structure, seismicity, active tectonics, topography and climate, has been frequently subjected to natural disasters resulting in great losses of life and property. Field studies and investigations of disasters indicate that large portions of the land surface, population and infrastructure of the region have been subjected to earthquakes in the past or will be subjected to earthquakes in the future. Knowing this, it has become more imperative to develop earthquake seismic vulnerability mapping and damage-control schemes targeting to reducing the degree of natural disaster damage to an economically manageable level.

The Jordan-Dead Sea Rift is considered the major fault system in the region as a result of motion between the Arabian and African plates. Seismicity information including historic and prehistoric data indicates that major destructive earthquakes have occurred in the Jordan - Dead Sea region, (*see* Figure 16.1). The instrumental seismicity of the region shows a concentration of earthquake activity along the major trend of the Rift and its associated zones. One of the recent hazard maps published by the Jordan Seismological Observatory (JSO) shows that the (format) area south

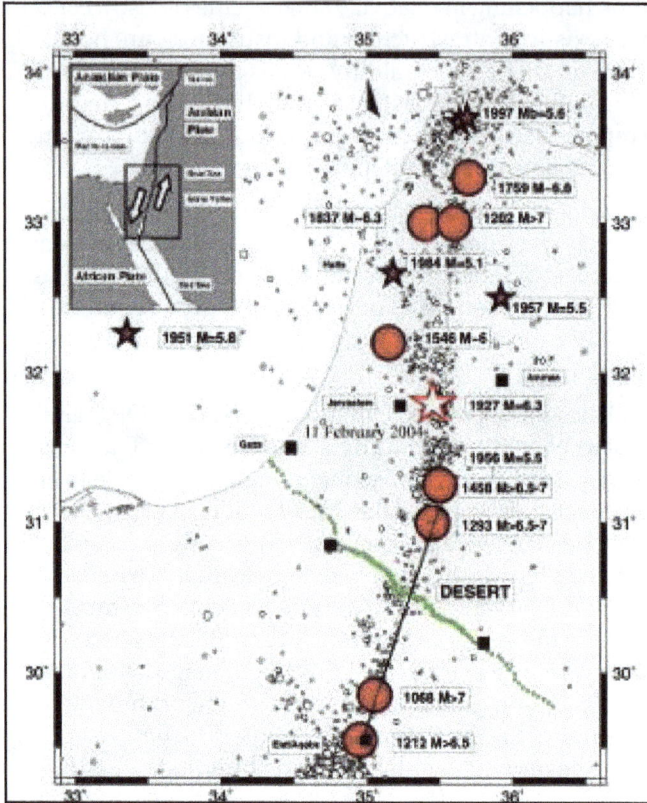

**Figure 16.1: Seismic Activity in the Dead Sea Transform Region;
The Map Shows Locations of Historical Earthquakes and the most Recent
Earthquake of 11 February 2004, ML 5.2.**

of the Dead Sea is of the largest seismic risk in comparison with the areas north and on its both sides.

An earthquake of magnitude greater than 6, Richter Scale, is expected to occur in the southern or in the northern parts of the Dead Sea during a return period of less than 100 years. Taking into consideration the major instrumental destructive earthquake of the region occurred in 1927 (6.25 magnitude with epicentre in northern part of the Dead Sea), a major destructive earthquake is expected any time in the near future and will be epicentered along the Dead Sea transform fault. The earthquake is expected to cause considerable damages and losses because of high vulnerability of common buildings.

Recent studies conducted by the Urban Planning and Disaster Risk Reduction Center (UPDRRC) at An- Najah National University estimated that a total damage of around 5 per cent to 15 per cent and partial damage of 25 per cent or more are expected in some areas (Jalal Al Dabbeek 2007, 2008 and 2010).

Based on the seismic peak ground acceleration map (PGA Map) for the region, the State of palestine has four seismic zones (*see* Figure 16.2).

The problems relating to earthquakes in Palestine can be summarized, but not limited, to the following:

☆ High vulnerability to earthquake damages and losses, as a direct result of very high percentages of weak buildings that do not comply with seismic resistance requirements.

☆ Lack of adequate national programs and public policies on preparedness, mitigation, and emergency response.

☆ Weak institutional capacity in disaster management and rescue operations.

☆ Weak of awareness by citizens, and weak capacity of professionals, engineers, and decision makers

☆ Weak of comprehensive, reliable, and easily accessible resources about seismic vulnerability of buildings which should be available for relevant Public/Governmental institutions to support decisions and long term urban plans.

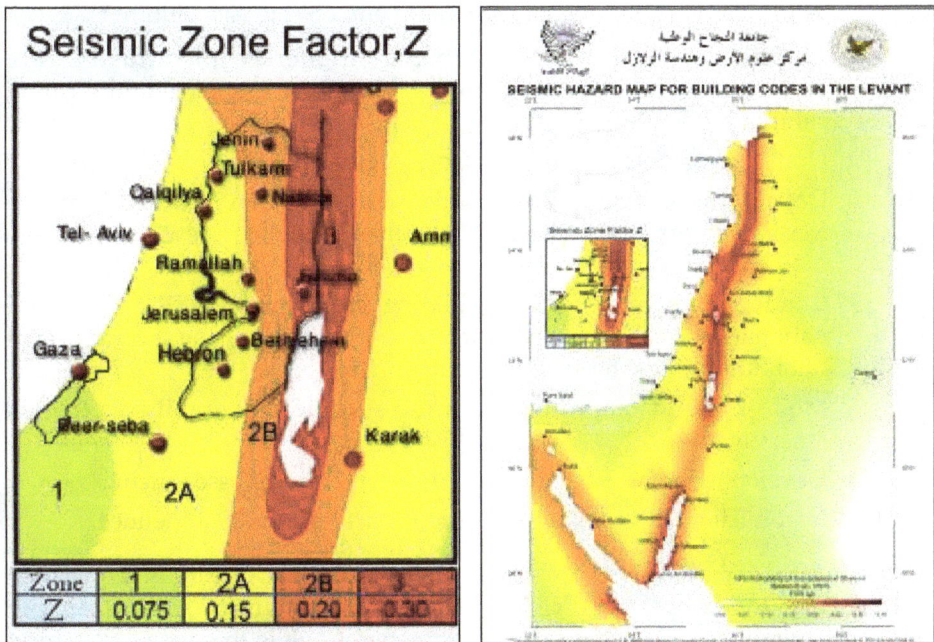

Figure 16.2: Seismic Hazard Map and Seismic Zone Factor.
(*Source:* ESSEU, Earth Sciences and Seismic Engineering Unit at NNU)

Academic Hubs: Using Science and Technology to Build Resilience

The main requirements and strategies needed to make scientific institutions work as academic hubs in DRR are: embedding DRR culturally, building civil societies capacity to respond to disaster in a targeted and DRR manner, becoming information-hub locally, nationally and regionally adopting:

- ☆ Target groups web-based platforms: individual citizens, professionals and decision-makers
- ☆ Local, national and international stakeholders' integrations and networking concepts;
- ☆ Prioritizing applied research and community-service programs;
- ☆ Holistic-approach concepts and methodologies.

Urban Planning and Disaster Risk Reduction Center (UPDRRC) at An-Najah National University brought players, using scientific knowledge and community services driving through change towards policy, preparedness and public awareness. UPDRRC, as an academic hub, has an important role in enhancing the Palestinian communities' disaster resilience, through adopting holistic approaches to DRR activities adopting scientific strategies driving together decision-makers, practitioners and the public towards sustainable risk reduction, exceeding traditional academic centers parameters.

As a result, this approach had a wide remit and objectives including:

- ☆ Assisting the government and practitioners with infrastructure vulnerability and local site effect conditions assessments and creating solutions;
- ☆ Drafting new Seismic Building Code regulations.
- ☆ Developing several courses and programmes on DRR, including a Master's program on Disaster Risk Management as a result of outcome of SASPARM1 project (SASPARM 1, 2012 - 2014).
- ☆ Conducting several capacity building programs on DRR at local and national levels, these included tens of training courses on seismic design of buildings, disaster risk management, *etc.*
- ☆ Modeling and mapping: Hazards, seismic vulnerability and risk assessment.
- ☆ Developing and conducting post disaster damages assessment.
- ☆ Introducing DRR requirements on physical planning guidelines.

In addition to the holistic approaches mentioned above, UPDRRC in cooperation with other units and centers at NNU developed and conducted several capacity building programs within civil society and general public to cope with natural disasters by using several dissemination activities which included:

- ☆ Developing engineering courses for non-engineers and urban planning courses for non-planners at NNU,
- ☆ Developing and conducting training courses on DRR for Journalists to create a common language between media and scientists and other stakeholders and target groups which are usually involved on DRR,

☆ Community service programs: each year 5000 students at NNU have community service course as a compulsory course. In this program, each student should do the following: Blood donation, 50 working hours with emergency response organizations (*see* Figure 16.3) and other organizations, working for/or with vulnerable citizens' organizations (with children's, mothers, handicapped or disabled persons, etc) and managing meetings and workshops in favor of these organizations. Parts of these students have the possibility to attend short courses on disaster risk management.

☆ Conducting hundreds of Meetings, Public Lectures, Workshops, Training Courses, *etc.* (*see* Figure 16.4, as well as UPDRRC, SASPARM1 and SASPARM2 web sites).

☆ Conducting several dissemination and public awareness activities by using all available media: TVs, Radios, newspapers, social media, *etc.*

☆ Developing and implementing a public seismic Poll to measure the seismic awareness levels of individual citizen in Palestine.

For more details about the above mentioned activities see the UPDRRC web site (**www.najah.edu/ar/community/scientific-centers/urban-planning-and-disaster-risk-reduction-center/),** SASPARM1 web site (www.sasparm.ps), and SASPARM2 web site (www.sasparm2.com).

Based on networking and integration concepts and to achieve its goals and objectives, UPDRRC participated in and/or conducted several local, regional and international DRR projects and activities in cooperation with UN organizations and other scientific research institutes, such as: EUCENTRE and IUSS in Italy, GFZ in Germany, USGS in USA, *etc.* The paper contains examples for several dissemination activities as well as a detailed case study of an important scientific project "Support Action for Strengthening Palestine capabilities for seismic Risk Mitigation - SASPARM2".

Support Action for Strengthening Palestine Capabilities for Seismic Risk Mitigaton (SASPARM 2 - Project)

Towards Integrated Seismic Risk Assessment in Palestine

Using large-scale seismic risk assessment studies for reduction of potential losses is becoming an ever more popular trend around the globe. Accordingly, a number of different models and techniques for the characterization of the different risk variables have proliferated in the recent years. Furthermore, the quality, or accuracy of risk estimates will be certainly higher when a truly integrative model is employed, characterizing hazard, (physical and social) vulnerability and exposure in the most complete as possible manner. Regions with a large percentage of non-seismically designed buildings are particularly vulnerable to seismic events and are those that can be the most to benefit from risk assessment studies for decision making. As such, the main purpose of this study is to propose a framework, based in Open Quake (an open-source software for seismic risk and hazard assessment

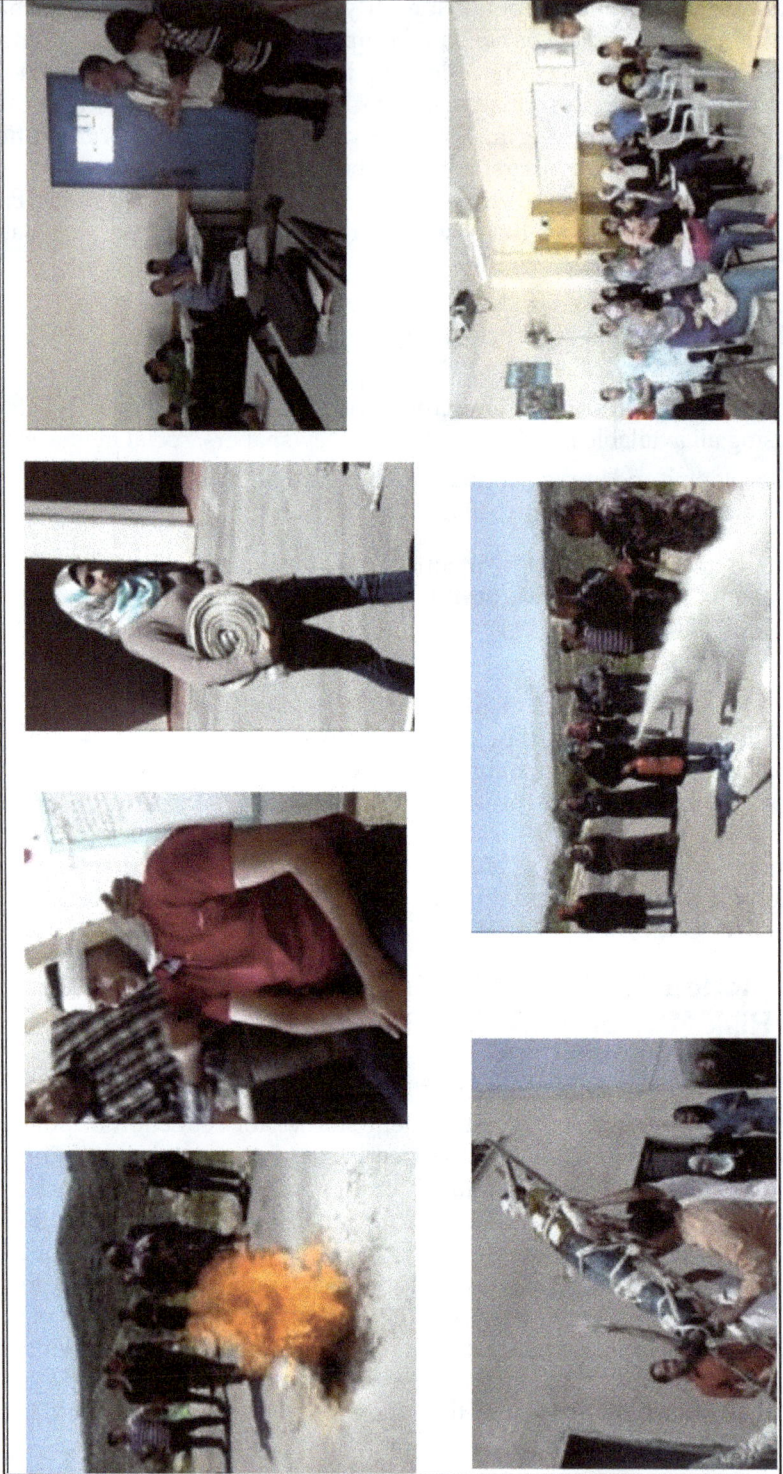

Figure 16.3: Several Activities from Community Service Courses/Programmes (for details: https://www.najah.edu/en/community/scientific-centers/).

Figure 16.4: Tenths of Meetings, Public Lectures, Workshops, Training Courses, *etc.*
(for more details see www.sasparm.ps).

developed within the global earthquake model (GEM) initiative for integrated seismic risk assessment in Palestine), where earthquake induced risk awareness is still at an early stage. A state-of-the-art hazard model as well as vulnerability and exposure models specifically built upon local field surveys and national data collection, are proposed. The outcome of the study will enable the identification of the regions that are more vulnerable to earthquakes and future rapid loss assessment at a regional scale.

Objectives of the Project

The mitigation of seismic risk, which comes from the convolution of hazard, exposure and vulnerability is the main project objective. Indeed, it is not possible to act on hazard and it is nearly impossible to act on exposure of existing cities (cities cannot be moved), but it is possible and dutiful to reduce vulnerability. To achieve this, one of the major issues to overcome is the lack of proper risk perception by citizens, making difficult to implement plans for seismic risk mitigation. This project refers to a population living in Palestine that, thanks to the exploitation and dissemination of the results of the SASPARM FP7-Project (www.sasparm.ps), has become aware of the concept of seismic risk. Furthermore, Palestinian stakeholders, governmental organizations (GOs) and non-governmental organizations (NGOs) Institutions, students and practitioners have shown a huge interest in the SASPARM activities and their outcomes. In addition, a new Seismic Building Code has been recently introduced in Palestine. Therefore, the awareness of the local community is of fundamental importance since the citizens have to monitor their buildings and be able to understand, with and, when feasible, without the advice of an expert, if their house can withstand an earthquake or if retrofitting is required by applying Seismic Standards. The practitioners as well as the GOs and NGOs stakeholders have to be aware of the importance of the right application and implementation of the new Seismic Building Code with the final goal of improving the seismic risk mitigation in Palestine with the support and collaboration of EU partners. Within this context, the need for prevention in the field of seismic risk is strongly required and this encouraged the proposal of the new project, named SASPARM 2.0. It represents a strategic avenue and a comprehensive advance of the post-SASPARM environment and involves the same Consortium – the An-Najah National University, Eucentre and the Institute for Advanced Studies (IUSS) of Pavia.

Actions and means Involved

The work of the project is split into 8 operative tasks, which operate within the framework of a Web-Based Platform (WBP) (task G) for seismic risk mitigation. The WBP integrates a database (DB) to collect vulnerability data on buildings (task B), self-assessment tools to allow common people and practitioners to understand potentially unsafe situations (task F) and how to mitigate the risk related to critical situations (task C). Training will play a fundamental role in the project activities and the target groups are students, practitioners, citizens and stakeholders (task D). The knowledge of seismic risk pending on individual properties will raise the awareness of citizens on the related seismic hazard and will help them to be addressed towards policy for risk management aimed to mitigate the socio-economic losses such as insurance coverage (task E). The WBP will constitute the tool linking the different actors and policies throughout the disaster management cycle, which translates into operational the technical data provided for the risk assessment to be used by the risk management governance. Nablus city will represent the real case study for the implementation and calibration of the project actions then this will be extended to the other Palestinian main cities through the direct involvement of the above mentioned targeted groups. The interaction between project tasks is shown in Figure 16.5. In addition to those shown in the plot, the tasks of project management (task A) and publicity (task H) will be implemented (For more details see www.sasparm2.com and/or www.sasparm.ps).

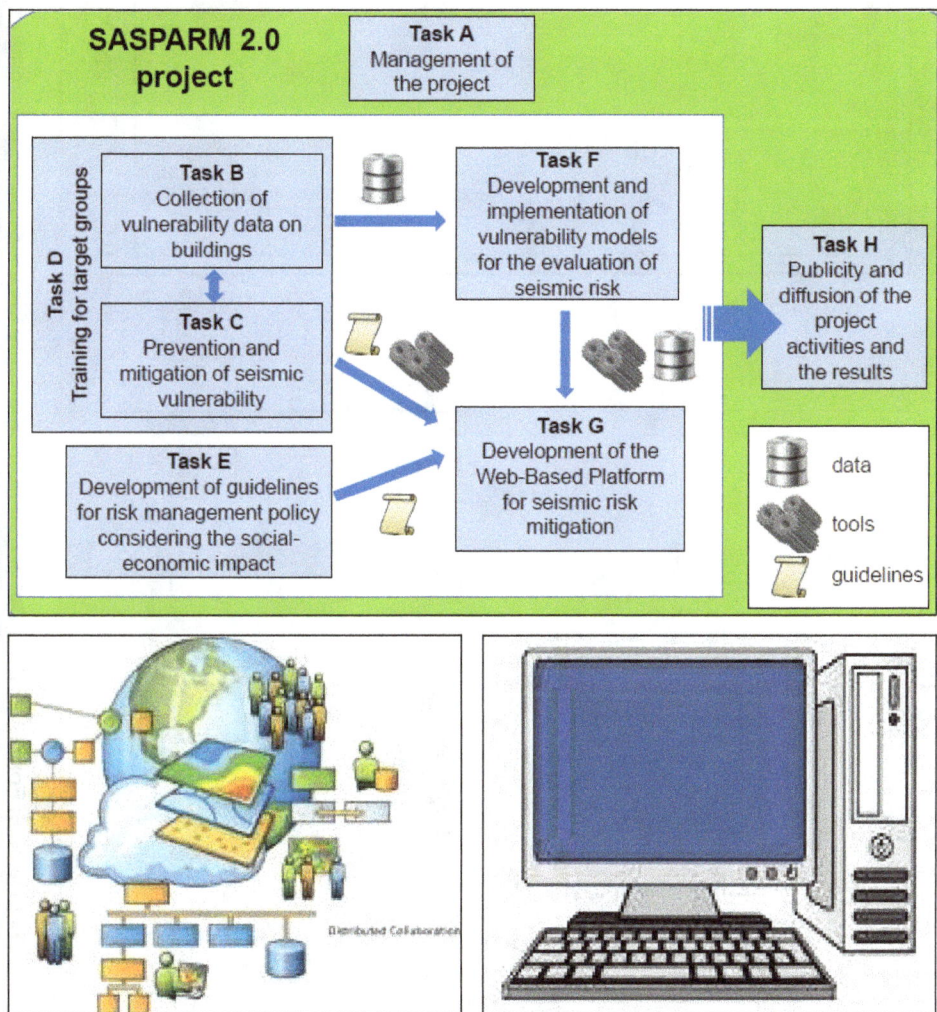

Figure 16.5: The Interaction between Project Tasks.

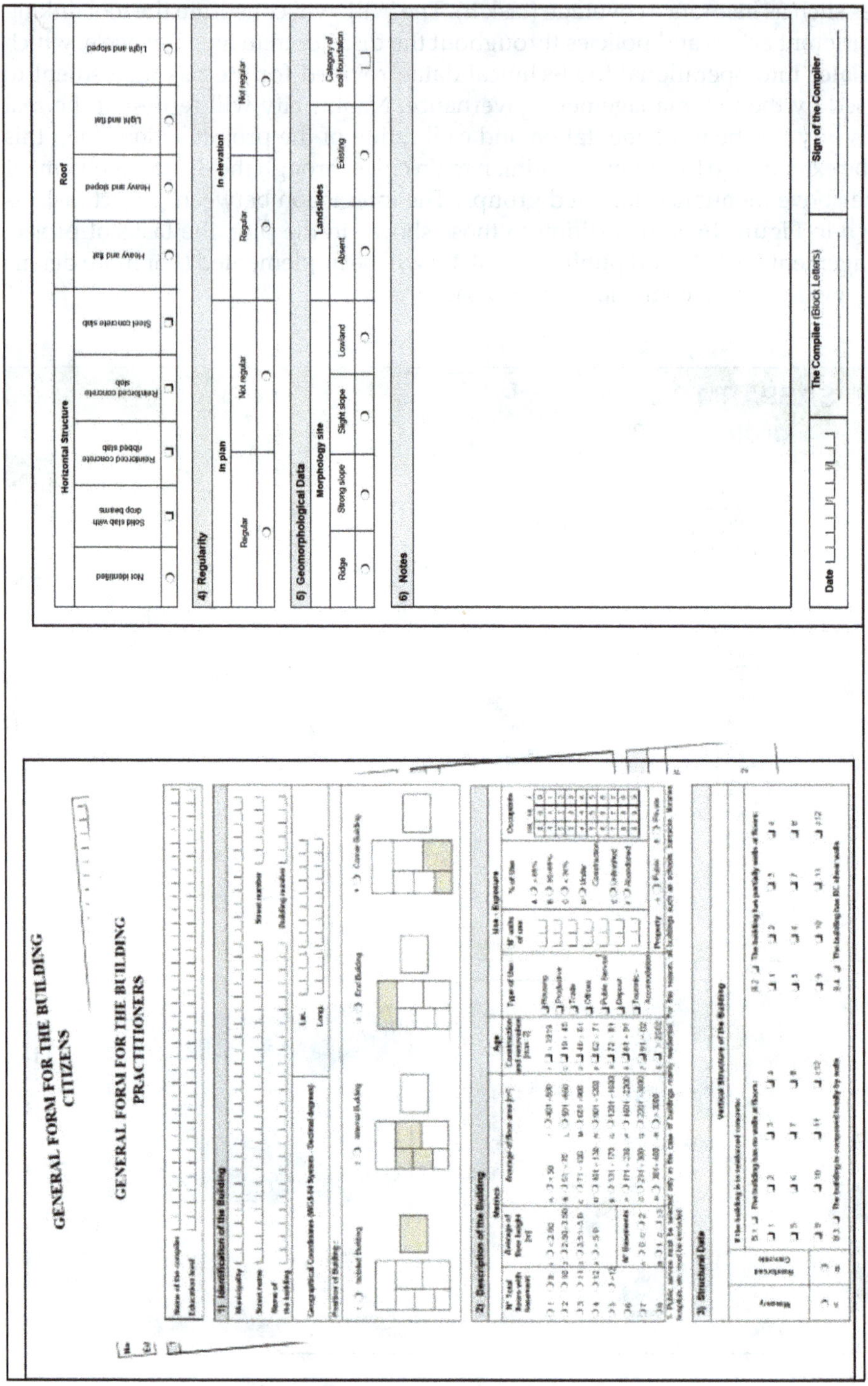

Figure 16.6: Forms for Data Collection (General forms for the buildings: Practitioners and Citizens forms).

Data Collection

The in situ building data collection will be done through forms by general citizens and practitioners. All information collected through the forms will be used to identify the vulnerability classes of the buildings according to their structural typology (*see* Figure 16.6). Appropriate retrofitting measures for the mitigation of seismic risk will be suggested to the end users of the platform.

Training and Workshops

To achieve the project goals and objectives, the project tasks have several training courses and workshops, such as:

☆ Training for university students

☆ Training for practitioner engineers

☆ Training for citizens

☆ Workshops and lecturers for stakeholders and policy makers

☆ Identify the vulnerability classes of the buildings according to their structural data

☆ Appropriate retrofitting measures for the mitigation of seismic risk will be suggested to the end users of the platform.

Expected Results

The main outcomes of the project will be:

☆ An increased awareness of seismic risk by the stakholders involved in the project: students, citizens, practitioners, GOs and NGOs stakeholders

☆ A shared DB, including a large number of vulnerability data

☆ A WBP that integrates the data above through vulnerability models developed for the

☆ Palestinian building typologies, to evaluate seismic risk

☆ Guidelines on the implementation of measures to reduce vulnerability hence mitigate seismic risk

☆ Guidelines for risk management policy towards mitigating the impact of socio-economic losses.

Projects Follow Up

☆ Extend the case study of Nablus municipality not only to all the other Palestinian municipalities but also to other Developing Countries and European Countries

☆ Engage policy makers and government to foster long-term actions. Moreover, promote Palestinian stakeholders' activities in a risk mitigation perspective with the foundation of a Palestinian Civil Protection Mechanism

☆ Establish the concepts of risk governance to account for the possibility of earthquake insurance coverage (considering that the related cost would be reduced if private initiative in retrofitting world is taken)

☆ Ensure the maintenance of the Web-Based Platform to collect larger amounts of data, first on seismic vulnerability of citizens' properties, and second on public buildings in order to keep the process of increasing awareness going on after the project lifetime

☆ Promote new undergraduate and graduate courses on seismic risk mitigation since the training on this topic will play a fundamental role for the continuation of the project goals even after its lifetime. For these reasons, additional resources will be found to organize a new Master's program in Palestine at the An-Najah National University, covering the topics of seismic risk mitigation. This action is already strongly encouraged by the Ministry of Education and higher Education.

References

1. Jalal Al Dabbeek, (2007). "Vulnerability, and expected seismic performance of buildings in West Bank, Palestine". *The Islamic University Journal (Series of Natural Studies and Engineering)*, Vol. 15, No. 1, 193-217.

2. Jalal Al Dabbeek and Radwan El Kelani, (2008). Rapid Assessment of Seismic Vulnerability in Palestinian Refugee Camps, *Journal of Applied Sciences, Journal of Applied Sciences* (8): 1371-1382, ISSN 1812-5624008 Asian Network for Scientific Information.

3. Jalal Al Dabbeek, (2010). An Assessment on Disaster Risk Reduction in the Occupied

4. Palestinian Territory (2010). *An - Najah Univ. J. Res. (N. Sc.)* Vol. 24, 2010.

5. Jardaneh, I., Al-Dabbeek J. and Al-Sahili K. (2013). Slope Instability of Nablus–Al-Bathan Road, to be published, *International Conference on Civil Engineering, ICCE*, 28-30, October, 2013, Al Bireh, Palesitne.

6. Jalal Al Dabbeek and others (2016). the implementation of the Sendai Framework for Disaster Reducation 2015-2030, UNISDR Science and Technology Conference, Switzerland 27-29 January 2016.

7. Making Cities Resilient Report, 2012. UNISDR.

8. Support Action for Strengthening Palestine capabilities for seismic Risk Mitigation (SASPARM 1 - Project). www.sasparm.ps

9. Support Action for Strengthening Palestine capabilities for seismic Risk Mitigation (SASPARM 2 - Project). www.sasparm2.com

10. Urban Planning and Disaster Risk Reduction at An Najah National University website www.najah.edu/ar/community/scientific-centers/urban-planning-and-disaster-risk-reduction-center/

Chapter 17

A Designing Experience of a National Research Agenda on Disaster Risk Reduction

Alejandro Linayo

Research Center on Disaster Risk Reduction CIGIR,
Foundation for Seismic Risk Prevention FUNDAPRIS,
Merida - Venezuela
E-mail: alejandrolinayo@gmail.com

ABSTRACT

The objective of this paper is to illustrate a designing and implementation experience of a National Research Agenda on Disaster Risk Reduction that was developed by the Ministry of Science and Technology of the Bolivarian Republic of Venezuela, between years 2000 and 2007. The goal of this applied research agenda was to promote a wider and deeper commitment of scientific and academic communities, with the diagnostic, analysis and solution of priority research problems that needed to be solved in order to improve national disaster risk reduction capacities.

The methodological framework used as part of the designing process of this Disaster Risk Reduction applied research agenda included the use of different consensus methodologies (Delphi, Snowball Sampling, Diffuse Logic, Soft System Thinking Methodologies, *etc.*). This effort was particularly useful for the definition of priority research areas and projects that finally were identified, promoted and developed with the participation and cooperation of a large number of experts.

The design and implementation of this research agenda was the result of a long and wide process of negotiation between multiple actors from diverse knowledge areas. This singular experience operated in Venezuela between years 2000 and 2007, and it has been recognized as a pioneer initiative to promote science and technology efforts in order to attend the complex challenges that must be faced to reduce integrally the risk of disaster in any country.

Keywords: Applied research, Science and technology, Disaster risk management.

Background

In 1999 the Government of Venezuela decided to create a Ministry of Science and Technology (MCT) in order to promote a deeper commitment from scientific and academic actors with the analysis and resolution of the most important problems of this South American country. The slogan of this new public institution was "Science and Technology for the People", and based on this political conception, the MCT started to work in the identification and design of research agendas focused in attend important social problems of this South American country.

Meanwhile the MCT was starting activities, in December of 1999 one of the worst disasters of in the last decades struck the north coast of this country, particularly in the territory of the state of Vargas, when torrential rains and severe mudslides killed thousands of people, destroying thousands of homes and collapsing almost all the infrastructure (potable water, roads, energy lines, phone lines, *etc.*) of the most affected areas. As result of this situation several coastal towns (Los Corales, Macuto, Mamo, El Piache, *etc.*) were buried under 1 to 3 meters (3.2 to 9.8 feet) of mud and a high percentage of homes were simply swept away to the ocean. Estimated damages calculated for this disaster were around$ 4 billion. It is estimated that more than 8,000 homes were destroyed, 75,000 people were displaced and that more than 60 kilometers (37 miles) of the coastline was significantly altered.

The simultaneous occurrence of the Disaster of December 1999 and the efforts for the consolidation of a new politic for scientific and technological development in the country, promoted that on year 2000 a recently created Ministry of Science and Technology launched an Applied Research National Program on Socio-natural Risk Management and Disaster Reduction, under the guidance and with the participation of some of the most experienced academics and specialists on diverse topics of disaster risk reduction in the country.

Two mayor goals were defined for this research program: The first one was the definition of better ways to deal with disaster preparation and response protocols, and the second, and most important one, was focused on the disaster risk reduction problematic and in the promotion of scientific and technological solutions to improve national risk prevention and mitigation capacities.

An element that was essential for the definition of this national research program was the recognition of the historical recurrence of different hazards and disasters situations that have affected this sur American country during its five centuries. An history in which could be found more than 200 strong earthquakes registered since the first one reported in 1530 (an event that destroyed the first Spanish buildings constructed in the city of Cumana), and that additionally include, only in the lapse from years 1900 to 2015, more than 1658 emergencies and disasters associated to landslides, 4398 floods and debrish flows associated to hydro climatic phenomenons and more than 3000 emergencies associated to technological and industrial accidents. All this statistics can be consulted in the online database of disasters http://www.estudiosydesastres.info.ve.

Previous conditions of social and institutional construction of disaster risk scenarios could be found in almost every disasters suffered either in Venezuela as in others Latin American countries. This fact invites us to think that disasters, far from being naturals, are always *socio-naturals*, and this is because there are always social processes of vulnerability construction behind any disaster. What could be eventually considered as natural is the event (in this case the strong rainfalls) that detonates unsustainable development models.

The most important conceptual elements assumed to define the working framework of this program were taken from a few national and regional scientific organizations, as the Foundation for Seismic Risk Prevention FUNDAPRIS, the Research Center on Disaster Risk Reduction CIGIR, and the Network for Social Studies on Disasters Prevention in Latin-American LaRED. For those three scientific organizations it was clear that, neither the disaster suffered in December 1999, nor other important disasters registered in Venezuela and in Latin America during last decades, could be referred as "natural disasters", and this position was sustained in the following elements:

It was possible to demonstrate that natural events that have triggered last important disasters in Venezuela during last century, far from being extraordinary or unexpected situations, constitute in fact inner manifestation of those territories dynamics. Specifically for the disaster of the state of Vargas, could be easily demonstrated that strong and long rainfalls (and its associated hydrogeological process) as the ones that stroke north central costs of Venezuela on December 1999 (Figure 17.1), are in fact events of well recognized historical recurrence of that Venezuelan region.

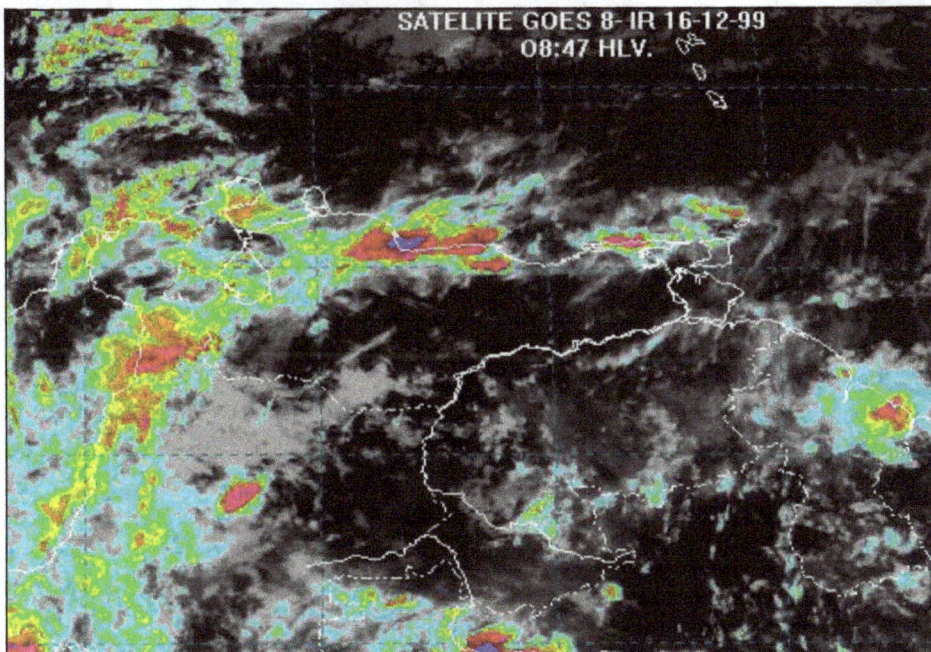

Figure 17.1: GOES Satellite Image (16/12/1999)
Showing Atmospheric Conditions during the Disaster.
That affected the central coast of Venezuela in December 1999.

There is a considerable amount of available information about previous strong rainfalls, landslides and debris flows similar very similar to those that affected Vargas 1999 (around 20 similar events can mention between years 1794 to 1951). It is also a fact that in those situations damages never reach the catastrophic proportions registered in December of 1999, but the main reason for this difference seems to be more associated with the very intense and anarchic process of urbanization registered in this region of the country between years 1960-2000, and with the way in which that urban occupation process creates perfect vulnerability conditions to register the levels of damages and suffering registered in 1999 (Figure 17.2).

Figure 17.2: Example of Damages Registered in the Disaster on Vargas (Linayo 2010).

Previous conditions of social and institutional construction of disaster risk scenarios could be found in almost every disasters suffered either in Venezuela as in others Latin American countries. This fact invites us to think that disasters, far from being naturals, are always *socio-naturals*, and this is because there are always social processes of vulnerability construction behind any disaster. What could be eventually considered as natural is the event (in this case the strong rainfalls) that detonates unsustainable development models (Figure 17.3).

Previous conditions of social and institutional construction of disaster risk scenarios could be found in almost every disasters suffered either in Venezuela as in others Latin American countries. This fact invites us to think that disasters, far from being naturals, are always *socio-naturals*, and this is because there are always social processes of vulnerability construction behind any disaster. What could be eventually considered as natural is the event (in this case the strong rainfalls) that detonates unsustainable development models (Figure 17.3).

Figure 17.3: Behind the Disaster of Vargas 1999, was Possible to Identify once again Important Mistakes and Lacks in Urban Planning and City Management (Linayo 2010).

Based on this approach, disasters could be understood as the natural consequence of non-managed risks, so to promote integral solutions for this problematic in our applied research program, it was decided to move from typical "disaster administration and response" approach, to a more integral disaster risk management approach were disaster risk management starts to be understood as condition for sustainable urban development.

Designing an Applied Research Agenda for Disaster Reduction

With the previous conceptual background established, a national program for disaster risk reduction started to work in Venezuela in March 2000. One of the first challenges faced for this was the need to offer answers to the following questions: ¿How to design a national agenda of applied research projects on disaster risk reduction, that includes all the knowledge involved on this complex area?, How to define priorities about which areas and/or which specific problems must be attended?, How to included the most important scientific and institutional actors in the designing and consolidation process of the agenda.

To answer those questions, one of the firsts and most important activities implemented was the designing of an agenda of priority research projects that were selected, promoted and executed, to improve institutional and social capacities on disaster risk reduction. This research agenda was the result of an 8 months process of negotiation between multiple academic, scientific and institutional actors from the most diverse knowledge. This process took place between January and September of year 2000 and as part of the activities implemented during this designing phase, a total of 52 workshops and meetings were celebrated, including conversations with more than 150 experts from different knowledge areas, and validation meetings with 63 public, institutional and social organizations, al this in an agenda that were developed in the most important cities of the country (Caracas, Mérida, Maracaibo, Cumana, Valencia, Barquisimeto and San Cristobal).

As part of the methodological approaches used during the designing process of the agenda, different consensus methodologies were used; particularly with the purpose of define levels of agreement on controversial aspects, subjects or points of view. From our experience, the most useful consensus strategies employed where:

- ✩ **Methodology Delphi:** It was used for being a systematic, interactive forecasting method which relies on a panel of experts. It suggest that the experts answer questionnaires in two or more rounds and after each round, a facilitator provides an anonymous summary of the experts' forecasts from the previous round as well as the reasons they provided for their judgments. Thus, experts are encouraged to revise their earlier answers in light of the replies of other members of their panel. In this way, during this process the range of the answers will decrease and the group will converge towards a "correct" answer (Hsu, 2007).

- ✩ **Methodology of Snowball Sampling:** Used as sampling technique to define the list of experts to be consulted for the agenda definition and where existing study subjects recruit future subjects from among their acquaintances. Thus the sample group is said to grow like a rolling snowball. In our case as the sample builded up, enough data was gathered to be useful for the final desing process (Goodman, 1961). This sampling technique is often used in hidden populations which are difficult for researchers to access (as experts that are part of disaster respond organizations, emergency organizations, *etc.*).

- ✩ **Soft System Thinking Methodologies:** In this case those tools were used as an approach suggested to be follow in the different research projects focussed in organizational and institutional process modeling, particularlly in those cases of organizational research that was related both for general problem solving related with disaster risk reduction and in the management of change that was suggested in some specific national development policies and institutions (Wilson, B. and van Haperen, K. 2015).

As result of this process was possible to identify 12 applied research thematic areas for the programme:

1. Education and Human Resources Training Aspects
2. Natural and Technological Hazards Characterization
3. Urban Vulnerability Assessment
4. Risk and Public Management
5. Life Lines Vulnerability Reduction

6. Technology for Disaster Preparation.
7. Medical and Psychological Aspects
8. Legal and Organizational Aspects
9. Social and Cultural Aspects
10. Risk Economy and Productive Process
11. Technology for Disaster Response Organizations
12. Technology for Housing and Services Rehabilitation

After the identification of those thematic areas, a number of 4 to 10 specific types of problems were defined for each one of the areas. In this case some complementary consensus methodologies were again used with the main purpose to define agreement on specific priority problems that were necessary to be attended.

About the Implemention and Results of the Program

Once defined the agenda of specific disaster reduction related problems to be attended by the program, a national call for proposals was made using most important national newspapers, in order to invite the scientific and academic community as well as other research and technological developers groups to prepare and send their proposals for the evaluation process established for this initiative. (Figure 17.4).

Two national calls were made during the time that this program works. The first one was in 2001 and a second one in 2003. As result of this process, the research program for disaster risk reduction received, evaluated and supported an important number of research projects proposals and activities focused on disaster risk reduction issues. Some numbers that could be mentioned about that process are here:

☆ Draft proposals received: 3512
☆ Formal projects received: 240
☆ Approved projects: 55
☆ Estimated investment: 4,5 millions $.

The applied research program on disaster risk reduction operated in Venezuela between years 2001 and 2007, year in which the Ministry of Science and Technology change its original orientation and was transformed in the Ministry of Science, Technology and Light Industries. With this transformation the original program disappeared, but even until today it is recognized as an opportunity to understand the complexity and wide range of challenges that must be considered by academic and scientific actors when integral disasters risk reductions initiatives are promoted.

Including Disaster Risk Reduction inside the University Academic System

A remarkable result of the national survey made on year 2000 to identify the priorities that need to be attended to deal with disaster risk reduction, was that it shows that the highest importance area to be attended was related with professional

Figure 17.4: Example of National Call for Proposals in National Newspapers of September 2001 (Liñayo - Estévez 2000).

and citizen education and training. This general opinion was common in every one of the multiple actors consulted and it made that the first priority of the 12 working areas included in the applied research agenda was called *Education and Training Aspects*. The list of specific key problems defined for this area included the following:

Area 1: Education and Training Aspects

1. Methodologies for sustainable incorporation of disaster risk management in the basic education official program.
2. Curricular strategies to promote academic professionalization of our Disaster Management National System.
3. Curricular strategies for the incorporation of DRR in university strategic programs.
4. Methodologies for measuring the impact of education/training programs on DRR
5. Technologies to promote e-learning related with local disaster risk management

It is important to note that national initiatives implemented to deal with those educative challenges in Venezuela, have been promoted years before the program for applied research on disaster risk reduction implemented in 2000. In fact in the next lines we will resume some results of a 35 years permanent learn-by-doing process, promoted specifically in the University of Los Andes (Venezuela) to incorporate disaster risk management in our superior education system.

Different Challenges - Different Educational Strategies

To understand the roll that our university should play in support of disaster risk reduction, and classify the kind of concrete activities that could be promoted by superior education systems in order to cooperate with the solution of this particular problem of unsustainable urban and rural development, we have suggested (Linayo 1997) that this goal demands to work in the three different main spaces of formal education that are resumed on the Table 17.1.

Table 17.1: Educational Strategies Proposed

Aspect	Objective	Educational Strategy
Professionalization of Emergency Management Systems	Guarantee effective, operative organizations required to deal with "non- wanted" but expected and ordinary urban emergencies.	Professional curriculum to certify professional capacity to work in first response organizations and emergency institutions.
Professionalization of Disaster Management Systems	Guarantee the social and institutional urban capacities required for a systemic urban response in case of disasters.	Professional curriculum to certify professional capacity to work in organizations for disaster management.
Incorporation of pertinent disasters risk reduction elements in strategic academic university programs	Guarantee an adequate disaster risk reduction capacity in each one of professionals that will work as urban development builders.	Transversal incorporation of disaster risk management elements on each one of "vulnerability builders" professional careers.

It is very important to differentiate clearly what is needed in each one of those three formative areas, and particularly to understand professional profile differences that exist between the Emergency Management, and the Disaster Management. About this could be said that we consider a mistake to think that an Urban Disaster can be assumed and handled as a "Big Emergency", so a professional profile to work in emergency management organizations is equivalent to the professional profile require to work on disaster response organizations. A Disaster, by definition, represents a situation that changes completely the institutional and social dynamic of the urban/rural affected context. In disasters scenarios the classical model of institutional and systemic response, usually leaded by fixed protocols that are applied to help "defenseless" citizens, simply disappear, and that is because a disaster definition imply a situation that exceeds institutional capacities of the emergency institutional systems, and demands the active participation of well and previously trained and coordinated social and institutional actors.

Based on these ideas, the university curriculum necessary for professional certification on disaster management must include, in addition to standard knowledge required to deal with emergency management operations, some curricular elements related with management process of extreme instability scenarios, in which, for example, non-linear behavior, self organization of complex systems, and others elements associated with the theory of chaos seems to be the rule.

An Academic Degree in Emergency Management and Disaster Reduction

With previous elements as a background, in the Technological University of the state of Mérida (Venezuela), was created the first Latin-American formal academic experience to certificate professionals as 'disasetrologists', if we may coin the term. First steps to create this program started in the early 1990s, however, it was not until 1997 when the first potential candidates to study this university career were accepted. This academic program certifies professional skills in three different specialties: Urban Operations, Industrial Operations, and Citizen Self-protection. At present, near 500 professionals have been graduated from this program, after finishing a 41 courses and seminars curriculum (Figure 17.5), that worth a total of 108 credit units, and that is organized into six 18-week semesters (three years).

Academic Efforts to Avoid the Social Construction of Disaster Scenarios

The academic challenge that demands disaster risk reduction, invite us to promote university college professions that consider their respective responsibilities in the construction of disaster risk scenarios, and this is by far a much more complex and difficult educative challenge. The key point in here is the recognition that disaster scenarios are almost always "constructed" by city planners, engineers, architects and other professionals that came from our universities, so the educational strategy required to solve this problem, in spite of creating a particular degree in "urban risk management", demand better ways of increasing disaster risk management capacities on those potential urban vulnerability builders that study in our university colleges.

One first initiative that was implemented to deal with this problem, was the creation in the University of Los Andes (located in the state of Merida – Venezuela, and academically ranked as the first university of that country in January 2016) of a post degree program on socio-natural and technological risk reduction. This postgraduate master program is offered since year 2010, to professionals of different knowledge areas, with the objective to bring them the academic skills needed to incorporate disaster risk reduction elements in their respective professional activities (Figure 17.6).

The described master program for disaster risk reduction has been very successful in terms of demand and applied research products, but in terms of scope and impact, it was for sure necessary to go beyond, looking for better ways to included, disaster risk reduction in each one of the academics university programs offered in the country. This goal has been clearly established by the United Nations Educational, Science and Culture Organization - UNESCO, as part of the objectives of the Education for Sustainable Development United Nations Decade 2005-2014

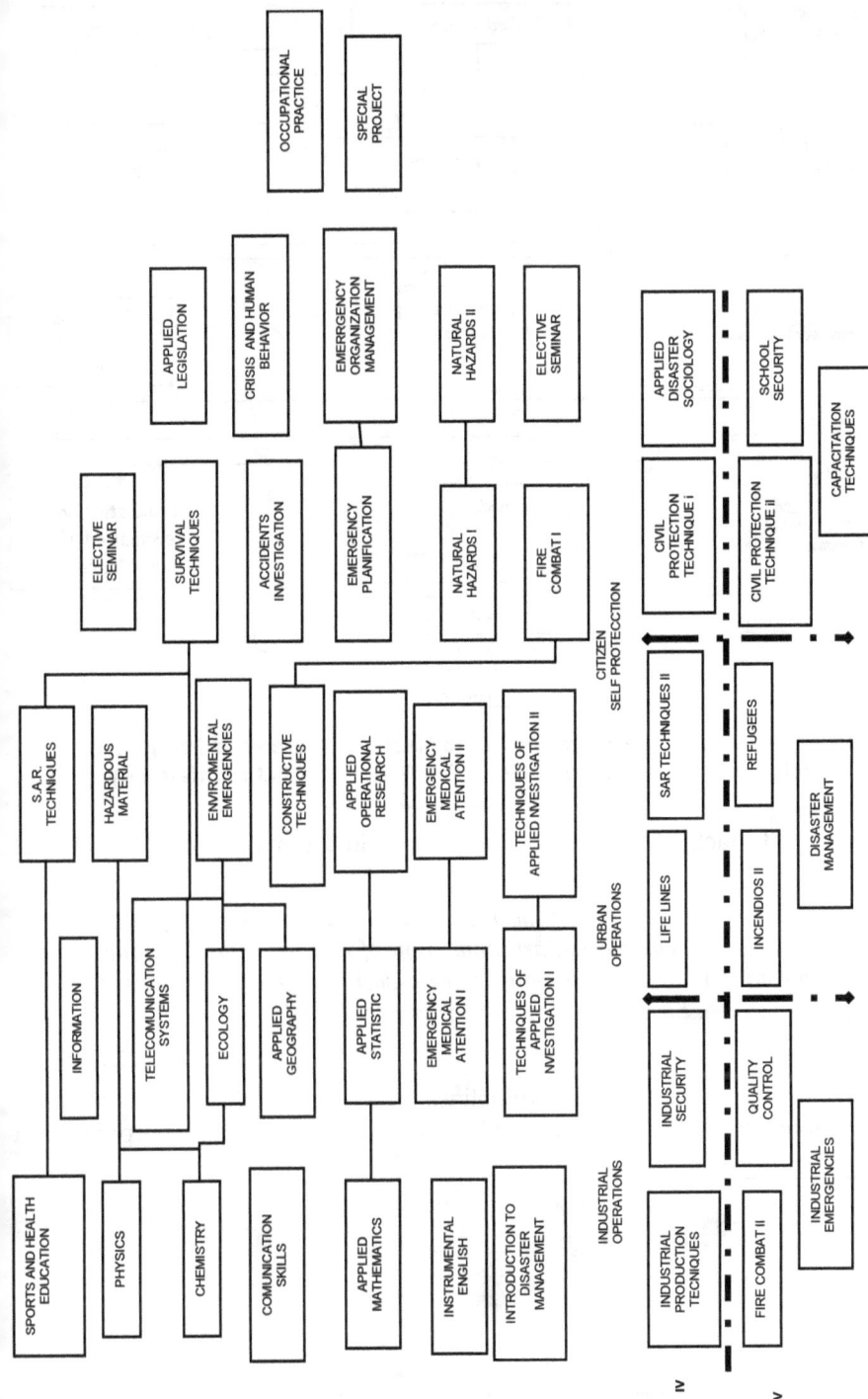

Figure 17.5: Professional Academic Program on Disaster Preparation and Response.

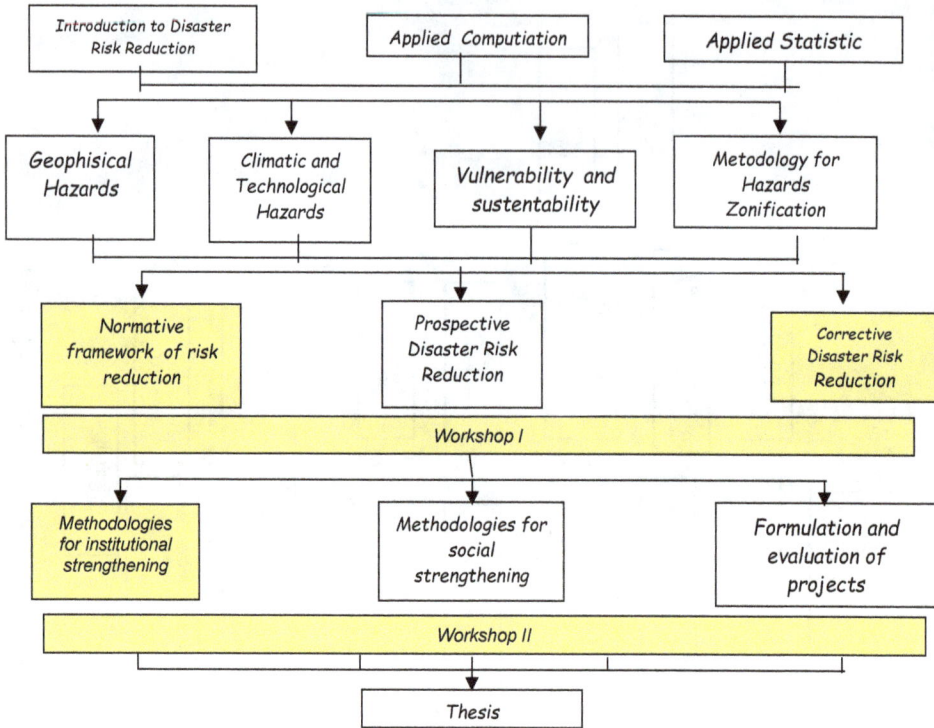

Figure 17.6: Curricular Structure of the Master Postgraduate Program on Socio-Natural and Technological Risk Reduction that is Offered by University of Los Andes.

(Figure 17.7). In fact, it is specifically expressed in the main documents of that initiative that:

> *"It is prioritary to promote the incorporation of disaster risk reduction as an indispensable and complementary dimention of any profetional profile, in order to asure in our college graduates the capacity to identify their specifics responsabilities in the construction of disaster risk scenarios."*

(ISDR-ONU, 2008)

Based on mentioned UNESCO guidelines, an educational research initiative has been promoted during last 7 years, in order to guaranty transversal incorporation of

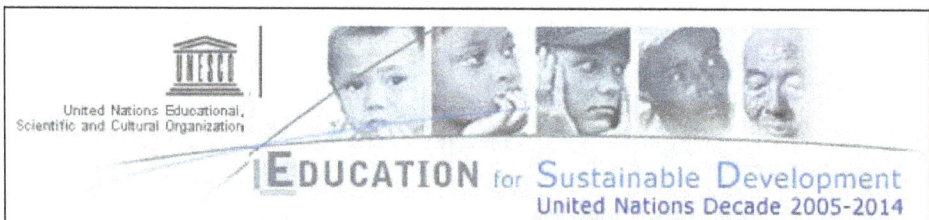

Figure 17.7: UNESCO Education for Sustainable Development United Nations Decade 2005-2014.

disaster risk management in priority university programs. The goal of this effort of the Doctorate on Educational Sciences of the University of Los Andes was to identify, in the whole list of professional degrees that are offered by national universities of Venezuela, the priority list of academic careers that must be reviewed and reinforced with specific and particular disaster risk reduction techniques.

For a quick resume of the methodology used to identify the top list of professional careers with high responsibility levels in the construction of disaster risk and vulnerability scenarios, we could be said that initially a national inventory of all university degrees programs offered in the country was made. After that a filtrate of all technical academic offers was done in order to keep only long-term university studies (minimum 4 years). Next a review of each one of program was made in order to put together professional academic degrees that could have different names but were very similar in curricular structure. After that, three numerical index were created to rank the of careers available, one associated with the social interest of the program, a second that measure the spread level of the program in the country and a third one related with the level of responsibility of their graduates in the building of disaster risk scenarios (based on experts opinion). Finally those parameters were statistically processed until a ranked list of priority academics careers to incorporate disaster risk reduction was identified.

Methodology and Basic Results for Proritary Programes Identification

1. Total number of university programs: 574
2. Total number of technical careers: 363
3. Filters applied to similar programs: 141
4. Definition of priority parameters:
    a. Index of social interest
    b. Index of coverage
    c. Index of DRR impact
        (a) Parameters normalization and ponderation
        (b) Priority list of programs to be reinforced
        (c) Top ten selection and pilots efforts of intervetion

Probably the most important to say about this effort, is that those results have started to be used by some national universities, to promote curricular reengineering process and even institutional commitmentsof universities colleges with disaster risk reduction. We will see in the future how far these steps contribute for a safer and sustainable society.

Conclusions

A previous requirement for promoting extensive intervention of any complex development problem, demands to know as much as possible the problematic situation and the conditions for each possible solutions. This happens as well for disaster risk management, and this fact justifies the promotion of national applied research agendas to strengthen capacities both for disaster risks prevention-mitigation, and for disaster preparation and response.

Usually research efforts conducted on this field seems to be isolated. Besides, their results barely correspond to the plans and projects of national development. Main efforts usually are focused in the rigorous characterization of threat sceneries (seismic, volcanic, meteorological threats) and physical vulnerability, aspects that are for sure very necessary, but that seems to insufficient to understand the complex challenges that disaster risk management demands.

The experience described here is an intent to go further in the identification and in the study of non- traditional (political, institutional, social, cultural, *etc.*) conditions that also define our vulnerary to disasters. Those aspects also deserves rigorous scientific research efforts that shows to our countries more efficient and appropriate politics to reduce integrally their disaster risk levels.

References

1. Hsu, C. (2007). "The Delphi Technique: Making Sense Of Consensus", available in the electronic journal "Practical Assesment, Research and Evaluation"; Vol 12 - Nro 10 available in http://essentialsofmedicine.com/sites/default/files/ Delphi per cent 20Technique_ per cent 20Making per cent 20Sense per cent 20Of per cent 20Consensus.pdf.

2. Goodman, L.A. (1961). "Snowball sampling". Annals of Mathematical Statistics. 32: 148–170.

3. Linayo, A., Massiani C. *et al.* (1993). 'An Academic Program in Security and Defense, Mention: Civil Protection', Technical Document, Ministerio de Educación, 26-45.

4. Linayo A. (1999). 'Curricular Design for a National Academic Degree on Disaster and Emergency Management', Technical Document, Ministerio de Educación Superior, 52-67.

5. Linayo A. (2006). 'National Strategies for Prevention and Mitigation of Disasters: Educative Aspects', Technical Document, Ministerio de Planificación, 78-96.

6. Liñayo, A. Estevez, R. (2003). "Bases para la formulación de Politicas Científicas en Reducción de Desastres". MCT.

7. Liñayo, A. (2002). "A systemic-interpretive approach to Disasters Administration in Latin America". Interpretive Sistemology Studies Center. University of Los Andes. Merida. 2001.

8. Maskrey A. (1993). "Disasters are not Natural". The Net - FLACSO. Editorial Tercer Mundo Bogotá.

9. Wilson, B. and van Haperen, K. (2015). Soft Systems Thinking, Methodology and the Management of Change, London: Palgrave MacMillan.

Colombo Resolution on Practical Guidelines to Minimize Hazards Due to Severe Natural Events

We, the participants of the International Workshop on "Mitigation of Disasters due to Severe Natural Events: From Policy to Practice" jointly organized by the National Science and Technology Commission (NASTEC), Sri Lanka and Centre for Science and Technology of the Non- Aligned and Other Developing Countries (NAM S&T Centre) in Colombo, Sri Lanka from 10th to 13th March 2016, respectively from Cambodia, Egypt, Hungary, India, Indonesia, Iran, Malaysia, Mauritius, Myanmar, Nepal, Nigeria, Pakistan, Palestine, Thailand, Uganda, Venezuela, Zambia, Zimbabwe and the host country Sri Lanka;

REALIZING THAT there is a significant number of severe natural events related human catastrophes in the world during the last few years of which the majority is in developing countries, both in urban and rural areas;

NOTING THAT there is loss of lives, high level of property and equipment damage and unexpected occurrences of service interruptions that make even vital public supply systems vulnerable to failure at critical circumstances which deprive the affected victims of basic needs for long durations and downtimes that may cause significant economic and social impacts at all levels;

RECOGNIZING THAT there is a lack of practical guidelines for the public to be adopted at grassroots level in the society and hierarchical order of risk reduction in many countries that leads to chaotic and haphazard decision making by the public under severe natural conditions, which enhances the level of human disaster;

REALIZING THAT there is a dearth of safety shelters, devices and equipment available and affordable to the public, especially in underprivileged communities, a fact that hinders the reduction of injuries and deaths due to severe events;

EMPHASIZING THE NEED for the development of feasible practical guidance for the public to be adopted in the event of severe natural situations, scientific and technological advancement in developing affordable safety systems, proper hierarchical order of safety reduction in the society, dissemination of knowledge and public awareness with respect to safety guidelines;

UNANIMOUSLY RESOLVE TO RECOMMEND the following 3-point roadmap for adoption by all concerned countries and parties and put into practice with immediate effect to minimize the losses of life, injuries, social chaos and economic losses under severe natural conditions:

1. The governmental authorities shall be informed and persuaded to:
   a. Collaborate with scientific and social service communities to obtain relevant data and information on the outcomes of pre-determined safety plans during severe natural events;
   b. Establish and/or strengthen early warning systems for severe natural events and disseminate relevant information to the public;
   c. Set forth safety guidelines to the public to be adopted under severe natural events;
   d. Define the hierarchical order of society in risk reduction due to severe natural event and enforce the societal order of command as a regulation;
   e. Periodically revalidate the effectiveness of guidelines at grassroots level, especially after a severe natural event in the respective country;
   f. Include essential concepts of severe natural event safety practices in school curricula;
   g. Encourage inventors and investors to develop affordable safety shelters, devices and equipment using new technologies, materials and tools for the public at all levels through promotion of business opportunities;
   h. Promote safety awareness among the public;
   i. Consider development of scientific and sociological tools in reduction of risk due to severe natural events as a priority area in government research funding programmes; and
   j. Focus on social capital for Community based Disaster Risk Reduction and Management (CBDRM) in formulation of policies.

2. The non-governmental organizations, private sector and other stakeholders shall be made aware of the risk of losses in the wake of a severe natural event and encouraged to:
   a. Organize educational and awareness programmes for different target groups regarding the practical guidelines to be adopted;
   b. Publish and share information on severe natural events, its hazards, vulnerability and risk precautions;
   c. Display proper safety instructions to be followed under severe natural

events, relevant to a particular region/location such as unstable hill slopes, floodplains, public gathering places at risk prone areas, *etc.*;

d. Enhance technical knowledge and skills among professionals having potential to develop sociological and technical tools to reduce effects of severe natural events; and

e. Encourage the public living in high risk areas to adopt risk reduction measures.

3. The academic and research communities shall be approached and advised to:

a. Develop research groups to conduct investigations on various aspects of severe natural events, including technological and sociological methods and tools;

b. Develop collaborative programmes and organize fora for sharing information and experience;

c. Facilitate professionals to get access to the up-to-date scientific and technical information through electronic media and other means;

d. Organize national/international training programmes and workshops with the support of the government, nongovernmental organizations and the private sector;

e. Provide advisory, consultancy and laboratory services to inventors and investors to develop commercially viable safety shelters, devices and equipment and test their products against national/international standards;

f. Validate/evaluate the effectiveness of the products, technologies and sociological tools under situations of severe natural events;

g. Organize training of trainers' programmes on severe natural event safety and protection; and

h. Develop synergy and devise innovations to address disaster risk management associated with severe natural events.

THUS ADOPTED AT COLOMBO, SRI LANKA THIS DAY, THE TWELFTH OF MARCH, AND TWO THOUSAND SIXTEEN.

Index